Ariacutty Jayendran
Rajah Jayendran

Englisch für Elektroniker

Literatur für das Grundstudium

Ariacutty Jayendran
Rajah Jayendran

Englisch für Elektroniker

Ein Lehr- und Übungsbuch für das technische Englisch

Mit 106 Abbildungen

Der Verlag Vieweg ist ein Unternehmen der Bertelsmann Fachinformation GmbH.

Umschlaggestaltung: Klaus Birk, Wiesbaden

Gedruckt auf säurefreiem Papier

ISBN-13: 978-3-528-03839-7 e-ISBN-13: 978-3-322-84907-6
DOI: 10.1007/978-3-322-84907-6

Vorwort

Dieses Lehrbuch wurde als Hilfe für Studenten und Ingenieure der Fachrichtung Elektronik geschrieben, um das für ihren Beruf notwendige Basiswissen der englischen Sprache zu erwerben. Obwohl bereits eine Vielzahl von Lehrbüchern über technisches Englisch erhältlich ist, existiert kein Buch zum Bereich Elektronik, dessen Konzeption gezielt für Studenten in den Anfangssemestern ihres Studiums an Universitäten und Fachhochschulen ausgelegt ist. Zudem kann das Buch durch Auslassen einiger mathematischer Teile auch an Technikerschulen, Fachschulen und Berufsschulen eingesetzt werden.

Das vorliegende Buch eignet sich als Grundlage für einen sechsmonatigen Kurs für Studenten der Fachrichtung Elektronik, die bereits über ein Grundwissen an Englisch verfügen; Schulenglisch sollte völlig ausreichend sein. Daher wird hier nicht versucht, Grammatik o.ä. zu thematisieren, und somit wendet sich das Buch nicht an absolute Anfänger. Wir haben uns jedoch bemüht, den Text einfach zu gestalten, indem wir auf kurze Sätze und die Vermeidung komplizierter Sprachkonstrukte bei der Sprachwahl achteten, um somit das Verständnis zu erhöhen.

Dieses Buch besteht aus 22 Lehreinheiten, wobei jede ein anderes Thema aus dem Bereich Elektronik behandelt. Jede Lehreinheit beinhaltet einen Textteil, der einen bestimmten Sachverhalt erläutert, an den sich ein Vokabelglossar sowie drei Übungen anschließen. Ein Glossar aller Vokabeln befindet sich am Ende des Buches.

Unter Elektronik versteht man ein weites Feld an Theorie und Anwendungen, weshalb man verständlicherweise in einem kleinem Buch wie diesem nur grundlegende Themen behandeln kann, die aber möglichst allen Spezialisierungsrichtungen als Basis dienen sollten. Naturgemäß ist die Themenauswahl ein Stück weit subjektiv, und andere Autoren mögen hierbei in Teilen anders entscheiden.

Die Themen, die wir ausgewählt haben, behandeln vier Bereiche:

a) Werkstoffe der Elektronik (Kapitel 1 & 2)
b) Halbleiterbauelemente (Kapitel 3 – 5, 14 & 16)
c) Analoge elektronische Schaltungen (Kapitel 6 – 13 & 15)
d) Digitale elektronische Schaltungen (Kapitel 17 – 22)

Bei der Behandlung der einzelnen Themen wurde versucht, nur wenig an mathematischen Formalismen einfließen zu lassen, jedoch wäre es unrealistisch und nicht wünschenswert, hierauf völlig zu verzichten.

Das vorliegende Buch eignet sich nicht nur für den Unterricht technischen Englischs, sondern insbesondere auch für das Selbststudium, da sich Musterlösungen zu allen Übungen am Ende des Buches befinden. Zudem bietet es die Gelegenheit, die grundlegenden Fachkenntnisse der Elektronik aufzufrischen, bzw. kennenzulernen.

Wir hoffen, mit unserem Buch Studenten, Ingenieuren, Technikern, etc. der Fachrichtung Elektronik einen leichteren Einstieg in das für sie relevante technische Englisch zu ermöglichen.

A. Jayendran
R. Jayendran

Contents

1 Conductors, insulators and semiconductors

Electrical materials are usually classified into three groups according to their electrical conductivity. Materials having a high electrical conductivity are termed good conductors and are usually metals. Materials which are poor conductors of electricity are called insulators. The third group consists of materials whose conductivities lie between those of metals and insulators. These materials which are called semiconductors play an important role in the field of electronics.

Energy band structure

The energies of electrons in a solid lie within certain restricted ranges called allowed energy bands. These allowed energy bands are separated by ranges of energy which the electrons cannot have called forbidden bands as shown in Fig 1.1.

Each energy band in a solid can only have a definite number of electrons. Some of the energy bands may be completely filled with electrons while others may only be partially filled with electrons. The band structure of a solid determines whether it behaves as a conductor, insulator, or semiconductor. A solid in which all the allowed bands are completely filled with electrons, behaves differently from a solid in which some bands are partially filled with electrons and some completely filled with electrons.

Completely and partially filled bands

When a band is completely filled with electrons, no net current is carried by the electrons in the band. Although each electron makes a contribution to the current, the sum of the contributions made by all the electrons is zero. Therefore the electrons in a completely filled band do not make a contribution to the conductivity of a solid. On the other hand electrons in a partially filled band can make a contribution to the conductivity of a solid.

Insulators

It has been stated that the electrons in a completely filled band do not make a contribution to the conductivity of a solid. It follows that a solid in which some bands are completely filled, and others completely empty, is an insulator.

Fig 1.2 shows the energy band structure of an insulator. The highest filled band is called the valence band and the next higher band the conduction band. The bandgap between the valence and conduction bands is large for an insulator. A good example of an insulator is diamond which has a bandgap of about 5 eV.

Semiconductors

At a temperature of 0 K a semiconductor behaves like an insulator, and has a band structure which is similar to that of an insulator. At higher temperatures however, a semiconductor has a conductivity which is higher than that of an insulator. The reason for this is that the bandgap of a semiconductor is smaller than that of an insulator being typically about 1 eV.

At absolute zero the valence band is completely filled, and the conduction band is completely empty. At room temperatures some of the electrons are transferred from the valence band into the conduction band. This results in the solid having a small but appreciable conductivity. As the temperature increases more electrons are transferred into the conduction band, and the conductivity increases. This is the opposite of what happens in the case of a metal. This increase in conductivity with temperature together with the small conductivity are the distinguishing features of a semiconductor. Fig 1.3 shows the energy band structure for a semiconductor at room temperature.

Metals

Elements which have only one valence electron are metals and good conductors of electricity. This is because the conduction band is only half-filled with electrons. Electrons which have two valence electrons can also behave like metals when their valence and conduction bands overlap. Elements of the alkaline earths like Calcium and Barium, have overlapping energy bands and behave like weak metals. Fig 1.4 and Fig 1.5 show the band structure of the two types of metal, one with a half filled conduction band, and the other with overlapping valence and conduction bands.

Vocabulary

absolute zero	absoluter Nullpunkt *m*	**identical**	identisch *adj*
allowed	erlaubt *adj*	**insulator**	Isolator *m*
appreciable	merklich, beträchtlich *adj*	**material**	Material *n*, Werkstoff *m*
behave	verhalten, sich benehmen *v*	**metal**	Metall *n*
band structure	Bandstruktur *f*	**overlap**	überlappen *v*
completely filled band	vollbesetztes Band *n*	**partially filled band**	teilbesetzes Band *n*
		range	Bereich *m*
conduction band	Leitungsband *n*	**restricted**	begrenzt *adj*
conductor	Leiter *m*	**role**	Rolle *f*
contribution	Beitrag *m*	**semiconductor**	Halbleiter *m*
current	Strom *m*	**separate**	trennen *v*
distinguish	sich unterscheiden *v*	**transfer**	übertragen *v*
energy bandgap	Energiebandabstand *m*	**valence band**	Valenzband *n*
field	Feld *n*	**valence electron**	Valenzelektron *n*
forbidden band	verbotenes Band *n*	**vicinity**	(nähere) Umgebung *f*

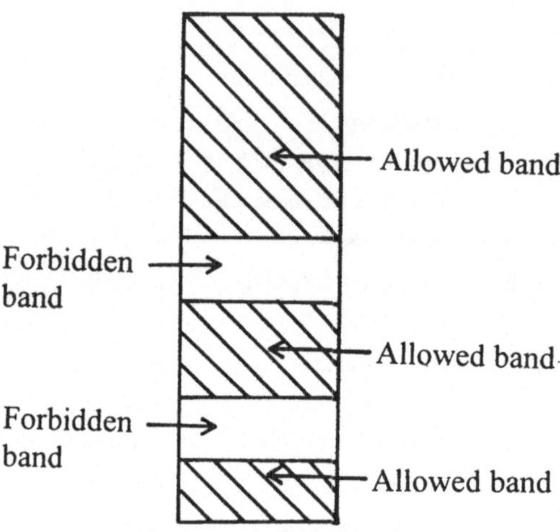

Fig 1.1 Allowed and forbidden energy bands in a solid

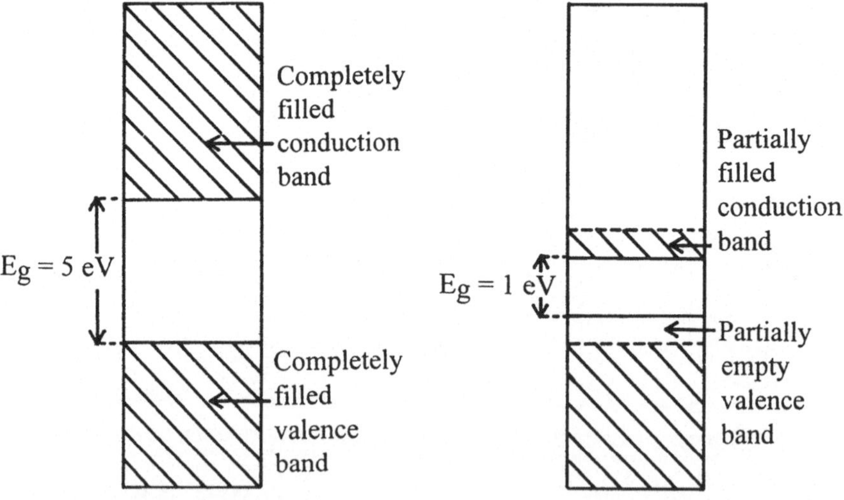

Fig 1.2 Energy band structure for an insulator

Fig 1.3 Energy band structure for a semiconductor at room temperature

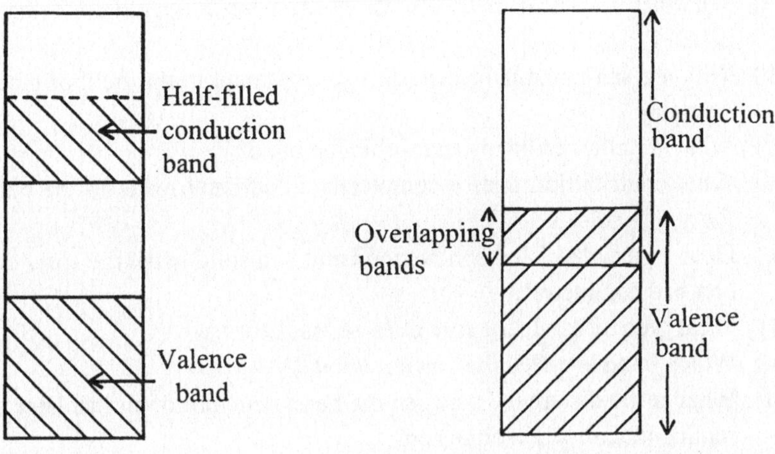

Fig 1.4 Energy band structure for a metal having one valence electron

Fig 1.5 Energy band structure for a metal having two valence electrons

Exercises I

1. Answer the following questions:

a) What are electrical materials which are poor conductors of electricity called ?
b) Why are semiconductor materials so important in the field of electronics today ?
c) What are allowed bands and forbidden bands ?
d) What contribution does a completely filled band make to the conductivity of a solid ?
e) Does a partially filled conduction band in a solid make the solid a good or a bad conductor ?
f) What type of band structure does an insulator have ?
g) Which band is called the valence band ?
h) What is the difference between the band structure of an insulator and a semiconductor at absolute zero ?
i) What are the distinguishing properties of a semiconductor ?
j) Why are elements which have only one valence electron good conductors of electricity ?

2.Fill in the gaps in the following sentences:

a) The electrical ____ of a solid depends on its ____ structure.
b) The ____ of the electrons in a solid can only lie within certain restricted ranges called ____.
c) Electrons in a ____ band do not make a contribution to the ____ of a solid.
d) A solid in which some bands are ____, and others ____ is an insulator.
e) The highest filled band is called the ____ band, and the ____ band is called the conduction band.
f) The ____ between the valence and conduction bands is large for ____.
g) At absolute zero, the ____ of a semiconductor is completely filled, and the ____ is completely empty.
h) Elements which have ____ electron are metals, because the conduction band is only ____.

i) Elements which have ____ electrons can also behave like metals, when their valence and conduction bands ____.
j) The increase in conductivity with ____ is one of the ____ of a semiconductor.

3. Translate into English:

a) Werkstoffe mit hoher elektrischer Leitfähigkeit bezeichnet man als gute Leiter. Werkstoffe, die schlechte elektrische Leitungseigenschaften besitzen, nennt man Isolatoren. Die Leitfähigkeit eines Halbleiters liegt zwischen der eines Leiters und der eines Isolators.
b) Die elektrische Leitfähigkeit eines Festkörpers hängt von seiner Energiebandstruktur ab. Die Energie der Elektronen in einem Festkörper kann nur in bestimmten Bereichen liegen, den erlaubten Bändern. Die erlaubten Bänder sind durch Bereiche getrennt, die von den Elektronen nicht eingenommen werden können, den verbotenen Bändern.
c) Das höchste gefüllte Band bezeichnet man als Valenzband und das nächsthöhere Band als Leitungsband. Der Bandabstand zwischen dem Valenz- und dem Leitungsband ist bei Isolatoren in der Größenordnung von 5 eV.
d) Bei einer Temperatur von 0 K ist die Bandstruktur eines Halbleiters identisch mit der eines Isolators, und er verhält sich wie ein Isolator. Bei Raumtemperatur werden Elektronen aus dem Valenzband in das Leitungsband übertragen.
e) Bei Temperaturerhöhung erhöht sich die Leitfähigkeit eines Halbleiters, was entgegengesetzt zum Verhalten eines Metalls ist. Diese Erhöhung der Leitfähigkeit bei Temperaturerhöhung ist eine der charakteristischen Eigenschaften eines Halbleiters.

2 Semiconductors

Intrinsic or pure semiconductors

Semiconductor materials used in electronic devices need to have a very high degree of purity and these highly purified semiconductors are called intrinsic semiconductors.

Many types of semiconductors are used in the fabrication of electronic devices. Important types are elements like silicon or germanium, and intermetallic compounds like GaAs or InSb. These materials need to be grown in the form of single crystals, before they can be used in electronic devices. However, polycrystalline materials like metallic oxides are also used in the fabrication of some electronic devices like for example thermistors.

Silicon which is probably the most widely used of the semiconductor materials, belongs to group IV of the periodic classification and crystallizes in the diamond structure. Each atom has four valence electrons and forms covalent bonds with four neighbouring atoms by a process of electron sharing. This results in a stable configuration in which each atom is surrounded by eight electrons, as shown in the two-dimensional diagram of Fig 2.1.

As the temperature increases, some electrons break away from a bond leaving a hole or vacancy behind as shown in Fig 2.2. Electrons and holes always occur in pairs in an intrinsic semiconductor, and therefore the concentration of electrons n is equal to the concentration of holes p (i.e. $n = p$ for a semiconductor).

In terms of the band theory, the breaking of bonds is equivalent to electrons being elevated into the conduction band leaving holes in the valence band as shown in Fig 2.3. When an electric field is applied, a current flows due to conduction by both electrons and holes.

Hole conduction

The mechanism of hole conduction when an electric field is applied may be understood from Fig 2.4. If there is a hole in position (1), then an electron (2) from an adjacent bond may drop into it, creating a vacancy at (2). This vacancy may be filled by an electron in position (3), creating a vacancy at (3). The position of a hole or vacancy can move in this way, and conduction can take place by a movement of holes which have a mobility μ_h.The conductivity of a semiconductor due to both electrons and holes is given by the expression

$$\sigma = n_e \mu_e e + p_e \mu_h e$$

Extrinsic or impurity semiconductors

n-type semiconductors

The usefulness of semiconductor materials increases considerably when small concentrations of impurity atoms are introduced into an intrinsic semiconductor. If atoms of elements having a valency of five like arsenic or phosphorous are introduced into a silicon crystal, each impurity atom forms covalent bonds with four neighbouring silicon atoms.The fifth electron belonging to the impurity atom cannot form a bond and remains only weakly attracted to the parent atom.

Impurity atoms of this type are easily ionized, the amount of energy required to remove the electron from the parent atom being of the order of 0.1 eV. Such impurity atoms are called donor atoms which when ionized give rise to fixed positive ions and mobile electrons without corresponding holes.

Impurity semiconductors of this type are known as n-type semiconductors, because the concentration of electrons is very much higher than the concentration of holes. Fig 2.5 shows a phosphorous impurity atom in a silicon crystal, and Fig 2.7 shows the band structure for an n-type crystal.The presence of fixed positive donor ions and mobile electrons is shown in Fig 2.9.

The conduction process in an n-type semiconductor is primarily due to electrons, and for this reason electrons are known as majority carriers, and holes as minority carriers.

p-type semiconductors

If we introduce impurity atoms having a valency of three like aluminium or boron into a silicon crystal, the situation is as shown in Fig 2.6. Only three of the four electrons required to form covalent bonds are now available. The impurity atom acquires an electron to form four covalent bonds and in this way becomes a negative ion. When an aluminium atom acquires an electron in this way, it becomes a negative aluminium ion. Impurity atoms of this type are called acceptor atoms, and when ionized give rise to fixed negative ions and mobile holes.

Impurity semiconductors of this type are known as p-type semiconductors, because the concentration of holes in them is much higher than the concentration of electrons. The conduction process is mainly due to holes, and holes are majority carriers while electrons are minority carriers. The band structure for a p-type semiconductor is shown in Fig 2.8, while Fig 2.10 shows a material containing fixed acceptor atoms and mobile holes.

It can be shown that the product of the electron and hole concentrations in a semiconductor at a given temperature remains constant and is given by the relation

$$np = n_i^2$$

where n_i is the concentration of holes or electrons in the intrinsic semiconductor.

Impurity level or doping

The introduction of impurities into a semiconductor is very often known as doping. Increasing the impurity level or doping by increasing the concentration of impurity atoms increases the majority carrier concentration and decreases the minority carrier concentration.

Vocabulary

adjacent	angrenzend *adj*	**intrinsic semiconductor**	Eigenhalbleiter *m*
configuration	Anordnung *f*	**ionize**	ionisieren *v*
covalent	kovalent *adj*	**majority carrier**	Majoritätsladungsträger *m*
create	erzeugen *v*	**minority carrier**	Minoritätsladungsträger *m*
device	Bauelement *n*	**mobility**	Beweglichkeit *f*
donor atom	Donatoratom *n*	**movement**	Bewegung *f*
doping	dotieren *v*	**neighbouring atom**	Nachbaratom *n*
elevate	erhöhen *v*	**n-type semiconductor**	n-Halbleiter *m*
equivalent	gleichwertig *adj*	**polycrystalline**	mehrkristallin *adj*
expression	Ausdruck *m*	**pure state**	reiner Zustand *m*
extrinsic semiconductor	dotierter Halbleiter *m*	**process**	Verfahren *n*
fabrication	Herstellung *f*	**p-type semiconductor**	p-Halbleiter *m*
hole or vacancy	Loch, Defektelektron *n*	**thermistor**	Heißleiter *m*
impurity	Verunreinigung *f*		
increase	zunehmen *v*		
intermetallic compound	intermetallische Verbindung *f*		

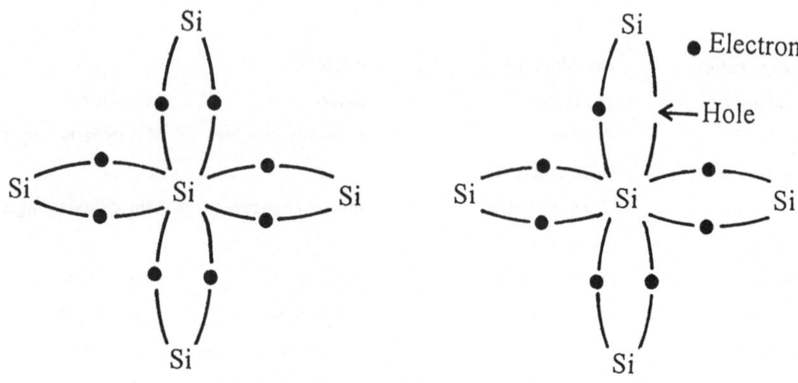

Fig 2.1 Si Atom surrounded by eight electrons in a silicon crystal

Fig 2.2 Formation of an electron and a hole in an intrinsic semiconductor

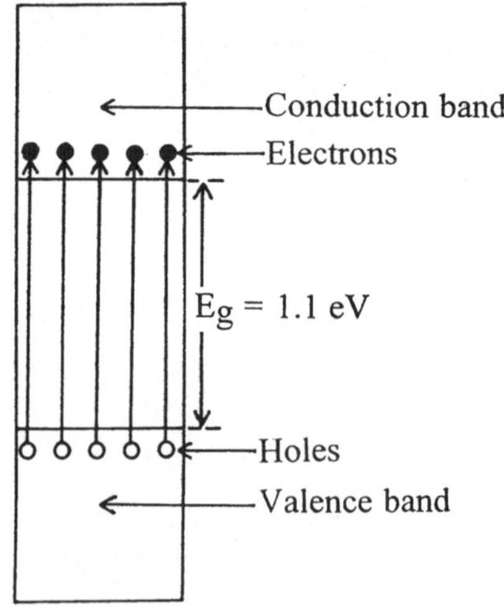

Fig 2.3 Electrons elevated into the conduction band leaving holes in the valence band

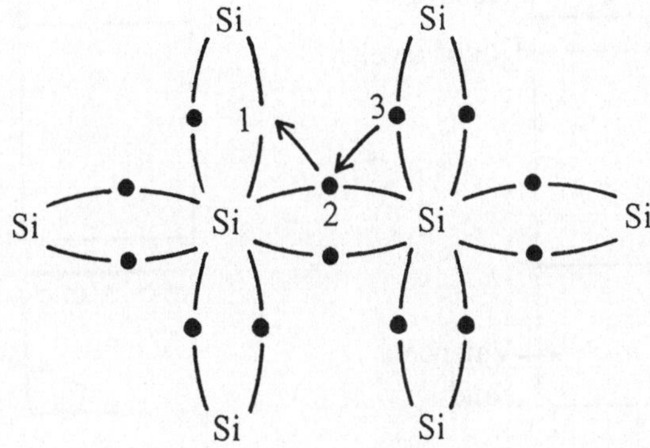

Fig 2.4 Movement of a hole in a silicon crystal

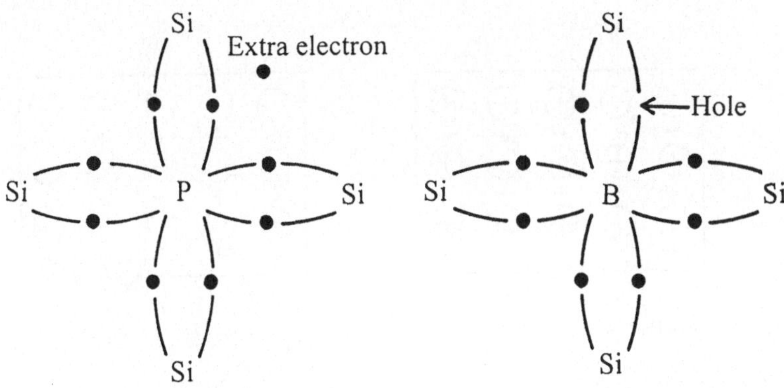

Fig 2.5 Creation of a mobile electron when a donor atom is introduced into a Silicon crystal

Fig 2.6 Creation of a hole when a boron acceptor atom is introduced into a Silicon crystal

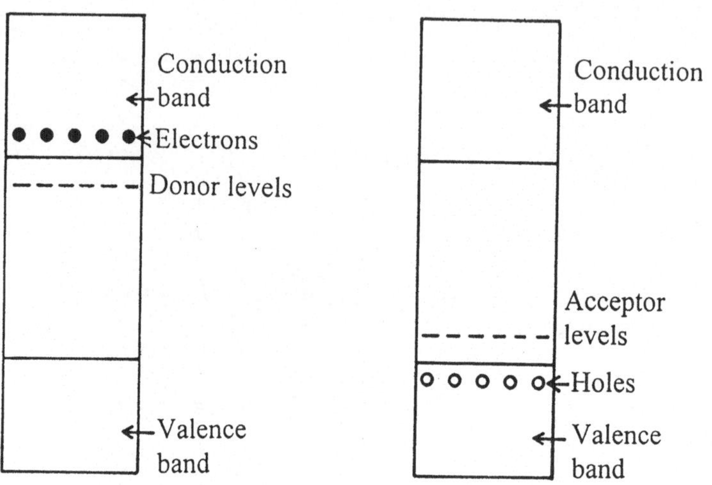

Fig 2.7 Energy band diagram for an n-type semiconductor

Fig 2.8 Energy band diagram for a p-type semiconductor

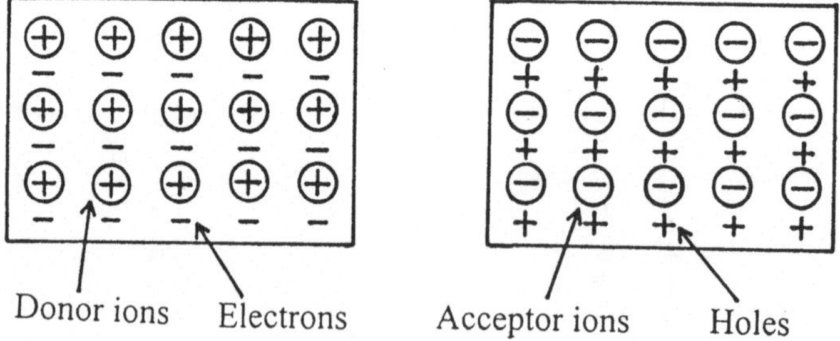

Fig 2.9 Fixed donor ions and mobile electrons in an n-type semiconductor

Fig 2.10 Fixed acceptor ions and mobile holes in a p-type semiconductor

Exercises II

1. Answer the following questions:

a) What are the other names given to pure and impure semiconductors ?
b) Give some examples of elements and compounds that are semiconductors.
c) What kinds of bonds exist in a silicon crystal and why are these bonds stable ?
d) Why do electrons and holes appear in pairs in a pure silicon crystal ?
e) What happens when an electric field is applied to a semiconductor ?
f) What types of impurity atoms are contained in an n-type semiconductor ?
g) What types of fixed and mobile charges exist in an n-type semiconductor ?
h) What is the conduction process in a p-type semiconductor primarily due to ?
i) What are the majority and minority carriers in a p-type semiconductor ?
j) Explain what happens to the carrier concentration when the level of doping in a semiconductor is increased.

2. Fill in the gaps in the following sentences:

a) Each silicon atom has four ____ and forms ____ with four neighbouring atoms.
b) Semiconductors in a highly ____ state are called ____ semiconductors.
c) Polycrystalline semiconductor materials are used in the ____ of some electronic devices, for example ____.
d) Silicon belongs to group IV of the ____ and crystallizes in the ____.
e) When an ____ is applied, a current flows due to conduction by both ____.
f) At higher temperatures electrons are ____ into the conduction band, leaving holes in the ____.
g) The ionization of donor atoms results in fixed ____ and mobile ____.
h) At a given temperature the ____ of the electrons and ____ remains constant.

i) The ____ of impurities into a semiconductor is very often known as ____.

j) When an aluminium atom ____ an electron in this way, it becomes a ____.

3. Translate into English:

a) Bei der Herstellung von elektronischen Bauelementen werden viele Arten von Halbleiterwerkstoffen verwendet. Einige von ihnen sind Elemente wie Silizium und Germanium, andere sind intermetallische Verbindungen wie Galliumarsenid (GaAs) oder Indiumantimonit (InSb).
b) Silizium gehört zur vierten Gruppe des Periodensystems und kristallisiert in der Diamantstruktur. Jedes Atom besitzt vier Valenzelektronen und geht kovalente Bindungen mit vier Nachbaratomen ein.
c) Störstellenatome, die bei Ionisierung die Bildung von stationären positiven Ionen und mobilen Elektronen verursachen, werden als Donatoratome bezeichnet. Halbleiter mit diesen Eigenschaften werden Halbleiter vom n-Typ genannt.
d) Der Einbau von Störstellen in Halbleiter wird häufig als dotieren bezeichnet. Eine Erhöhung der Dotierung erhöht die Konzentration der Majoritätsträger und verringert die Konzentration der Minoritätsträger.
e) Der Leitungsvorgang in einem Halbleiter vom n-Typ beruht hauptsächlich auf Elektronen. Aus diesem Grund werden die Elektronen Majoritätsträger und die Löcher Minoritätsträger genannt.

3 The junction diode

A p-n junction can be formed in a semiconductor crystal by introducing donor impurities into one part of the crystal and acceptor impurities into the other part of the crystal. A junction is formed in the region where the two types of impurities meet. A junction diode is constructed by attaching two ohmic metallic contacts to the ends of a semiconductor crystal which has a sharp p-n junction inside it. Such a device behaves like a rectifier allowing a large current to flow in one direction, and a negligible current to flow in the other direction.

The open circuited p-n junction

When nothing is connected to the ends of a diode, it is said to be on open circuit. Under open circuit conditions a potential difference is produced across the junction. The origin of the potential difference may be explained with the help of the diagram shown in Fig 3.1. Initially there is a high concentration of holes to the left of the junction, and a high concentration of electrons to the right of the junction. Since there is a density gradient, diffusion takes place across the junction holes moving to the right and electrons to the left.

The electrons and holes combine with each other, leaving negative acceptor ions on the left of the junction and positive donor ions on the right of the junction as shown in Fig 3.1(a). The region on either side of the junction which is depleted of electrons and holes is called a depletion or space charge region.

The variation of space charge across the junction is shown in Fig 3.1(b), the variation of field intensity in Fig 3.1(c), and the variation of potential in Fig 3.1(d). The charges on either side of the junction form a potential barrier, which opposes the further diffusion of electrons and holes across the junction, causing a state of equilibrium to be reached.

Application of external voltages

Reverse bias

If an external battery is connected to the terminals of a p-n diode, the potential barrier across the depletion layer is either increased or decreased depending on the polarity of the external voltage or bias. The polarity for reverse bias is shown in Fig 3.2. The negative terminal of the battery is connected to the p-side of the junction, and the positive terminal to the n-side of the junction. This increases the potential barrier across the junction causing the width of the depletion layer to increase as shown in Fig 3.2. In theory no current should flow across the junction, but in practice a very small current flows.

This is due to the fact that although the flow of majority carriers is opposed by the potential barrier, minority carriers can flow down the potential hill and are not influenced by the height of the barrier. This leads to the flow of a very small current I_S which is independent of the applied voltage.

Forward bias

The polarity for forward bias is shown in Fig 3.3. The positive terminal of the battery is connected to the p-side of the junction, and the negative terminal to the n-side of the junction. This reduces the potential barrier across the junction, and also the width of the depletion layer.

The equilibrium that was initially established between the diffusion of majority carriers and the opposing influence of the potential barrier at the junction is now disturbed. As a result of the lowering of the potential barrier, more holes cross the junction from right to left and similarly more electrons cross the junction in the reverse direction. This results in the continous flow of a large current.

Holes crossing from right to left are equivalent to electrons crossing from left to right. Hence the total current across the junction is the sum of the hole and electron currents.

The voltage-current characteristic

It has been stated that a very small current flows through a junction diode when it is reverse biased. This is called the reverse saturation current, and it is usually represented by the symbol I_S.

It can be shown that the forward current I is related to the external voltage by the relation

$$I = I_S\left(e^{eV/kT} - 1\right)$$

Fig 3.4 shows the voltage-current characteristics of a junction diode.

Vocabulary

attach	befestigen, anbringen *v*	**influence**	Einfluß *m*
avalanche	Lawine *f*	**junction**	Übergang *m*
characteristic curve	Kennlinie *f*	**junction diode**	Flächendiode *f*
contact	Anschluß *m*	**negligible**	vernachlässigbar *adj*
cross	kreuzen *v*	**open circuit**	Leerlauf *m*
density gradient	Dichtegradient *m*	**polarity**	Polarität *f*
depletion region	Raumladungszone *f*	**potential barrier**	Potentialbarriere *f*
device	Bauelement *n*	**potential difference**	Potentialdifferenz *f*
diffusion	Diffusion *f*	**rectifier**	Gleichrichter *m*
disturb	stören *v*	**reverse bias**	Sperrvorspannung *f*
equilibrium	Gleichgewicht *n*	**saturation**	Sättigung *f*
establish	herstellen *v*	**space charge region**	Raumladungszone *f*
external voltage	äußere Spannung *f*	**variation**	Änderung *f*
forward bias	Vorwärtsvorspannung *f*	**work function**	Austrittsarbeit *f*
forward current	Durchlaßstrom *m*		

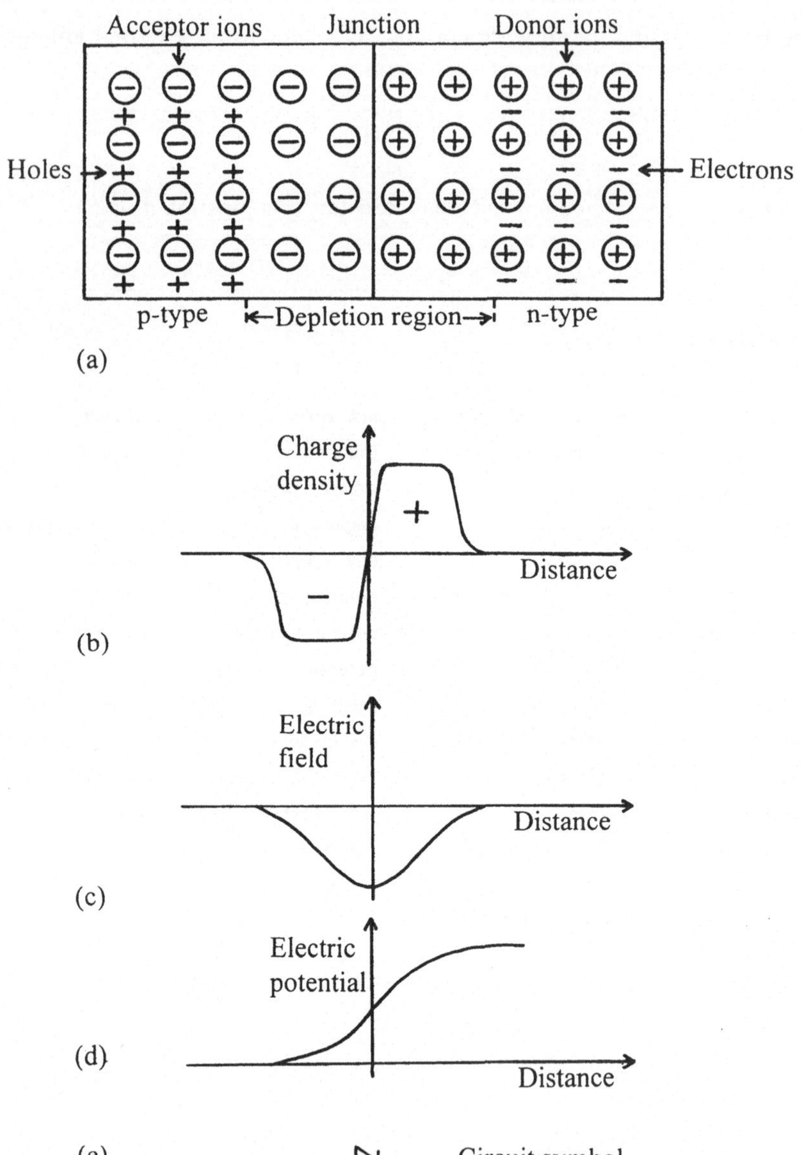

Fig 3.1 Open circuited p-n junction showing the charge density, electric field and the potential variation at the junction

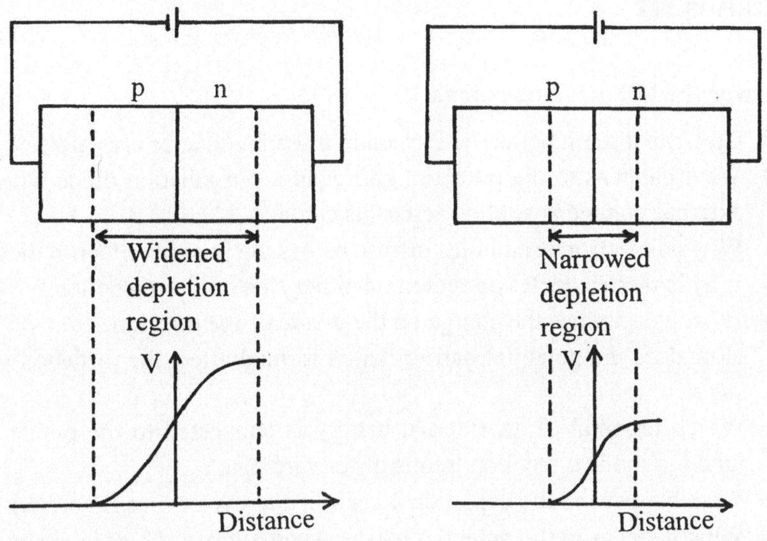

Fig 3.2 A p-n junction with reverse bias

Fig 3.3 A p-n junction with forward bias

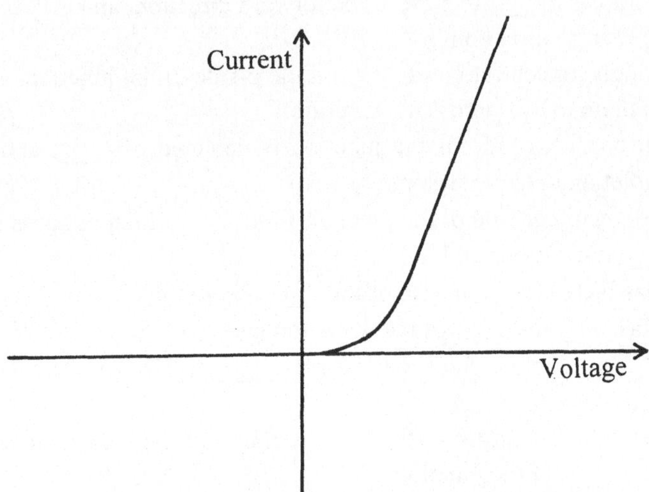

Fig 3.4 Variation of forward current with external voltage for a p-n junction diode

Exercises III

1. Answer the following questions:

a) How can a p-n junction be formed in a semiconductor crystal ?
b) What happens to the potential barrier in a p-n junction diode when an external voltage is applied across its contacts ?
c) Why do electrons and holes diffuse across the junction of a p-n diode ?
d) Why is there a depletion region on either side of a p-n junction ?
e) What polarity has the charge on the p-side of the junction ?
f) How does the potential barrier that is formed affect the diffusion process ?
g) Which terminal of an external battery is connected to the p-side of a junction diode under conditions of forward bias ?
h) What happens to the depletion layer when a reverse bias is applied ?
i) What happens to the potential barrier when a forward bias is applied ?
j) How are the hole and electron currents related to the total current ?

2. Fill in the gaps in the following sentences:

a) A rectifier allows a ____ to flow in the forward direction, and a ____ to flow in the reverse direction.
b) There is a high concentration of ____ on the p-side of the junction, and this causes them to ___ across the junction.
c) The region on either side of the junction is depleted of ____, and is called a depletion or ____ region.
d) The charges on either side of the junction form ____ which opposes the further ____ of electrons and holes.
e) Reverse bias increases the height of the ____ across the ____.
f) Forward bias ____ the ____ of the depletion layer.
g) Holes ____ the junction from the left are ____ to electrons crossing from the right.
h) Under reverse bias conditions, the ____ terminal of the battery is connected to the ____ of the junction.
i) The potential barrier ____ across the junction ____ the further diffusion of electrons and holes.

j) The electrons and holes ____ with each other leaving ____ on the left of the junction.

3. Translate into English:

a) In einem Halbleiterkristall kann ein pn-Übergang gebildet werden, indem auf der einen Seite des Kristalls Donatorstörstellen und auf der anderen Seite Akzeptorstörstellen eingebaut werden.
b) Eine Flächendiode ist ein Bauelement, das externe Metallkontakte besitzt, die auf den Enden eines Halbleiterkristalls mit scharfem pn-Übergang angebracht sind. Ein solches Bauelement verhält sich wie ein Gleichrichter, in dem es einen hohen Stromfluß in eine Richtung und einen vernachlässigbaren Stromfluß in die Gegenrichtung zuläßt.
c) Ursprünglich gibt es eine hohe Konzentration von Löchern auf der einen Seite des Übergangs und eine hohe Konzentration von Elektronen auf der anderen Seite. Dieser Dichtegradient veranlaßt die Löcher und Elektronen, den Übergang zu überqueren und miteinander zu kombinieren.
d) Eine Sperrvorspannung erhöht die Potentialbarriere am Übergang und auch die Breite der Raumladungszone. Eine Vorwärtsvorspannung andererseits verringert die Potentialbarriere und die Breite der Raumladungszone.
e) Wird eine Vorwärtsvorspannung angelegt, so wird das Gleichgewicht gestört, das zwischen der Diffusion von Majoritätsträgern und dem entgegenwirkenden Einfluß der Potentialbarriere bestand.

4 Other types of semiconductor diodes

Metal semiconductor diodes

The junction formed between a metal and a semiconductor can be either an ohmic junction or a rectifying junction. The leads or external contacts of a semiconductor device must under normal conditions form ohmic or nonrectifying junctions with the semiconductor. The junction between a metal and a semiconductor inside a metal-semiconductor diode must however, be a rectifying one. Such a diode is often called a Schottky diode. The external voltage-current characteristic curve of a Schottky diode is similar to that of a junction diode, but the physical mechanisms involved in the conduction process are more complicated.

Fig 4.1(a) shows a Schottky diode formed by the use of IC techniques. In this case, an n-type silicon semiconductor is used with aluminium as the metal to form a Schottky diode. The aluminium contact 1 is a rectifying contact, while contact 2 is an ohmic contact. Aluminium has a valency of 3, and acts as a p-type impurity when in contact with silicon. Since the effective work function of the semiconductor is less than that of the metal, electrons flow from the semiconductor into the metal. The metal acquires a negative charge, while the semiconductor acquires a positive charge.

In the forward direction, electrons from the semiconductor flow into the metal where electrons are plentiful. The Schottky diode is therefore a majority carrier device, as compared with the junction diode which is a minority carrier device.

Switching times for junction and Schottky diodes

In a junction diode, a delay is experienced in switching it from the forward to the reverse direction, because minority carriers stored near the junction must be removed first. The time required to remove the stored minority carriers is called the storage time t_s . The storage time for a Schottky diode is negligible, because current flow is mainly due to majority carriers. Elimination of storage time enables faster switching times to be achieved in Schottky diodes and

Schottky transistors. Metal-semiconductor point contact diodes can be used at microwave frequencies because their junction capacitances are very small.

Varactor diodes

It has been seen that when a p-n junction diode is reverse biased, electrons and holes do not cross the junction but move away from the junction. This results in positive donor ions and negative acceptor ions existing on either side of the junction. Further, the width of the depletion layer formed increases with increasing reverse voltage.The diode behaves like a capacitor whose capacitance varies with applied voltage. The incremental capacitance of the diode is given by the expression

$$C_T = \left|\frac{dQ}{dV}\right|$$

This voltage-dependent capacitance has typically values between 2 pF and 10 pF. One application for such a capacitance is in automatic frequency control circuits at VHF and UHF frequencies.

Avalanche and zener breakdown effects

When the reverse voltage applied to a junction diode is increased beyond a certain value, the current suddenly increases and the diode is said to be in the breakdown region. Two mechanisms of diode breakdown are recognized, avalanche breakdown and zener breakdown. In avalanche breakdown, thermally generated carriers which are part of the reverse saturation current fall down the junction potential barrier and acquire energy from the applied voltage source. These carriers have enough energy to disrupt the covalent bonds and form new electron-hole pairs. In this way there is a rapid multiplication in the number of charge carriers and the current in the reverse direction is greatly increased. This process is called avalanche breakdown.

Breakdown is also possible even when thermally generated charge carriers do not acquire enough energy to cause avalanche breakdown. This process known as zener breakdown occurs when a strong electric field exists at a junction. The strong field causes electrons to be removed from a covalent bond creating electron-hole pairs and causing an increase in the reverse current.

Zener diodes

The field intensity increases as the doping level is increased, and zener breakdown can occur for heavily doped diodes at voltages below 6V. The breakdown voltage is higher for lightly doped diodes, and in this case the breakdown is mainly due to avalanche multiplication. The breakdown characteristic of a zener diode is shown in Fig 4.2.

Tunnel diodes

The width of the depletion layer in a junction diode depends on the impurity concentration. In a tunnel diode, the impurity concentration is increased until it is about one part in a thousand. Under these conditions, the width of the potential barrier is only about one-fiftieth of the wavelength of light. When the widths are so small, quantum mechanical tunnelling can take place and an electron has a large probability of tunnelling or penetrating through the barrier.

Diodes of this type known as tunnel diodes have a volt-ampere characteristic which is very different from that of a p-n junction diode as shown in Fig 4.3. It will be seen that a large current flows in the reverse direction and that the curve has a negative resistance region in the forward direction.

Vocabulary

acquire	erwerben *v*	**incremental**	differentiell, zuwachsend *adj*
attach	anbringen, befestigen *v*	**lead**	Anschluß *m*, Zuleitung *f*
avalanche breakdown	Lawinendurchbruch *m*	**mechanism**	Mechanismus *m*
capacitance	Kapazität *f*	**ohmic**	ohmsch *adj*
complicated	kompliziert *adj*	**plentiful**	im Überfluß *adj*
construction	Aufbau *m*	**process**	Verfahren *n*
eliminate	beseitigen *v*	**storage time**	Speicherzeit *f*
enable	ermöglichen *v*	**store**	speichern *v*
experience	Erfahrung *f*	**rectify**	gleichrichten *v*
external contact	Außenanschluß *m*	**varactor diode**	Kapazitätsdiode *f*
field intensity	Feldstärke f	**zener breakdown**	Zenerdurchbruch *m*

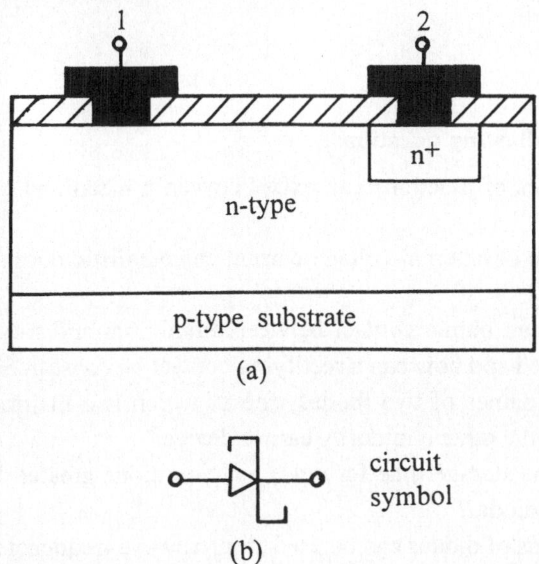

Fig 4.1 (a) A Schottky diode formed by using IC techniques, (b) the symbol for a Schottky diode

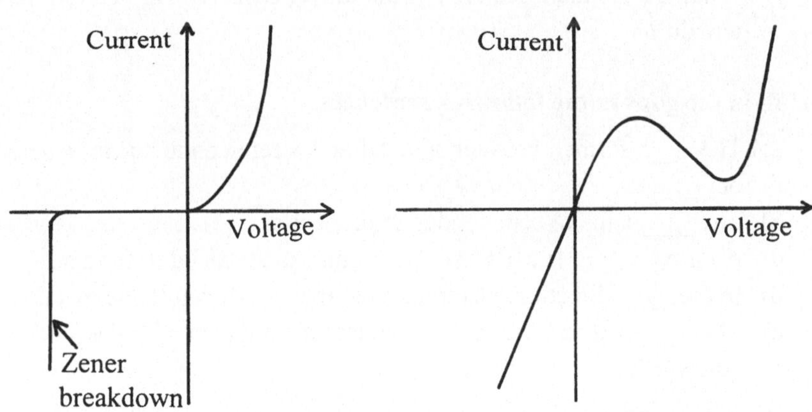

Fig 4.2 Breakdown characteristic for a zener diode

Fig 4.3 Voltage-current characteristic for a tunnel diode

Exercises IV

1. Answer the following questions:

a) What types of junctions can exist between a metal and a semiconductor?
b) What kind of external voltage-current characteristic does a Schottky diode have ?
c) How can an ohmic contact between aluminium and n-type silicon be constructed and how can a rectifying contact be constructed ?
d) Give the names of two diodes, one of which is a majority carrier device, and the other a minority carrier device.
e) Why is the storage time for a p-n junction diode greater than that for a Schottky diode ?
f) What kinds of diodes can be used at microwave frequencies ?
g) What happens to the width of the depletion layer in a junction diode, when the reverse voltage is increased ?
h) What are the conditions under which zener breakdown occurs ?
i) Give one example of a circuit in which varactor diodes can be used.
j) What are the main features of the tunnel diode voltage-current characteristic ?

2. Fill in the gaps in the following sentences:

a) The ____ formed between a metal and a semiconductor may be ohmic or ____.
b) The ____ a metal-semiconductor diode is a ____ .
c) An n-type ____ is used with aluminium as the metal to form a ____.
d) In the ____ direction, electrons from the ____ flow into the metal.
e) The ____ is a majority carrier device as compared with the junction diode which is a ____ device.
f) The storage time for a Schottky diode is ____ because the ____ is mainly due to majority carriers.
g) If a p-n junction is ____, electrons and ____ move away from the junction.
h) The width of the ____ increases with increasing ____.

i) The process known as ____ breakdown occurs when a strong ____ exists at a junction.

j) A ____ diode shows a negative resistance region in the ____.

3. Translate into English:

a) Der Übergang zwischen einem Metall und einem Halbleiter kann entweder ohmsch oder gleichrichtend sein. Der Aufbau von Metall-Halbleiter Dioden ist nur möglich, wenn gleichrichtende Kontakte am Übergang verwendet werden.

b) In einer Flächendiode tritt eine Verzögerung auf, wenn von der Vorwärtsrichtung auf die Sperrichtung umgeschaltet wird, da die nahe dem Übergang gespeicherten Minoritätsträger zunächst entfernt werden müssen. Die Zeit, die zum Entfernen der Minoritätsträger benötigt wird, wird Speicherzeit genannt.

c) Wenn eine pn-Diode in Sperrichtung vorgespannt wird, überqueren Elektronen und Löcher die Barriere nicht, sondern bewegen sich von dem Übergang weg. Eine Erhöhung des Sperrvorspannung vergrößert die Raumladungszone.

d) Wenn die an eine Flächendiode angelegte Spannung in Sperrichtung einen bestimmten Betrag überschreitet, steigt der Strom plötzlich an und die Diode befindet sich im sogenannten Durchbruchbereich. Es werden zwei Durchbruchmechanismen unterschieden, der Lawinendurchbruch und der Zenerdurchbruch.

e) Die Breite der Raumladungszone in einer Flächendiode hängt von der Störstellenkonzentration ab. Bei einer Tunneldiode wird die Störstellenkonzentration soweit erhöht, bis sie ungefähr 1/1000 beträgt.

5 The junction transistor

The junction transistor consists of three semiconductor regions with two sharp junctions between them. The regions are arranged in the order npn or pnp, and are called emitter, base, and collector. The two possible types of transistor are shown in Fig 5.1 with their circuit symbols. The central region is very narrow in width, and has a much lower impurity concentration than the two outer regions.

If no external voltages are applied the transistor is on open circuit, and under equilibrium conditions there is no net flow of electrons or holes across the two junctions. In a transistor that is symmetrical, the barrier heights at the two junctions are equal as shown in Fig 5.2(a) for a pnp transistor.

When a transistor is used as an amplifier, the emitter-base junction is forward biased, while the collector-base junction is reverse biased. This changes the heights of the potential barriers as shown in Fig 5.2(b). The potential barrier at the first junction is lowered, while the barrier at the second junction is increased in height. The circuit of an amplifier with a resistive load R_L and the appropriate biasing voltages is shown in Fig 5.2(b).

Under these conditions currents flow through the transistor and the external circuits. The external currents are the emitter current I_E, the collector current I_C, and the base current I_B.The main movements of charge carriers and their contributions to the external currents may be explained in terms of Fig 5.3. The main physical processes involved may be divided into three groups.

1. Holes in the emitter region move across the first junction into the base region, because the forward bias lowers the potential barrier across the junction below the equilibrium value. Most of these holes drift through the narrow base region without combining with electrons in the base, because the base has a low electron concentration. When the holes reach the second junction, the field is in such a direction as to accelerate them across the junction. These holes drift through the collector region, and are collected by the collector electrode. This gives rise to a collector current I_C. This

movement of holes from emitter to collector without recombination is shown in the first five lines of the diagram.

2. A small number of the holes that arrive in the base region combine with electrons. These electrons are replaced by new electrons which flow out of the base electrode. This flow of electrons out of the base electrode constitutes part of the base current, as shown in line 6 of the diagram.

3. The field at the emitter-base junction is in such a direction as to accelerate electrons in the base towards the emitter. This results in a flow of electrons from the base electrode into the emitter electrode. The flow of electrons constitutes part of the base current as shown in line 7 of the diagram.

Transistors are designed so as to make the base current small in comparison to the emitter current. This is achieved as has been stated before, by having a base region of narrow width and low impurity concentration. Most of the holes flow from the emitter to the collector without combining with the electrons in the base region. This keeps the base current I_B small.

$$I_B \ll I_E$$
$$I_E \approx I_C$$

From Kirchoff s laws,

$$I_E + I_B + I_C = 0$$

Leakage currents

When the emitter circuit is open, no emitter current flows but a small collector current flows. This small collector current called I_{C0} is the reverse saturation current of the reverse biased collector base junction, and is extremely temperature sensitive.

Current gain

The ratios of the external currents are important parameters in studying the behaviour of a transistor. The ratio of the collector to the emitter current is often represented by the large signal current gain parameter α where

$$\alpha = -\frac{I_C - I_{C0}}{I_E}$$

Typical values of α lie between 0.90 and 0.99.

The expression for I_C may be written as

$$I_C = -\alpha I_E + I_{C0}$$

If β is defined by

$$\beta = \frac{\alpha}{1-\alpha}$$

Then I_C becomes

$$I_C = (1+\beta) I_{C0} + \beta I_B$$

Since $I_B >> I_{C0}$

$$\beta \approx \frac{I_C}{I_B}$$

which is the ratio of the collector to the base current.

Small signal current gains

The incremental or small signal current gain parameters α' and β' may be defined as

$$\alpha' = \left.\frac{\Delta I_C}{\Delta I_E}\right|_{V_{CB}}$$

$$\beta' = \left.\frac{\Delta I_C}{\Delta I_B}\right|_{V_{CE}}$$

Vocabulary

accelerate	beschleunigen *v*	**drift**	sich verschieben, treiben *v*
achieve	erreichen *v*	**electrode**	Elektrode *f*
appropriate	passend, angemessen *adj*	**emitter**	Emitter *m*
base	Basis *f*	**explain**	erklären *v*
collector	Kollektor *m*	**involve**	umfassen, verwickeln *v*
combine	verbinden *v*	**load**	Last *f*
comparison	Vergleich *m*	**possible**	möglich *adj*
constitute	bilden *v*	**parameter**	Kenngröße *f*, Parameter *m*
contribution	Beitrag *n*	**region**	Bereich *m*, Region *f*
current gain	Stromverstärkung *f*	**recombination**	Wiedervereinigung *f*
design	entwerfen *v*	**symmetrical**	symmetrisch *adj*
direction	Richtung *f*	**width**	Breite *f*

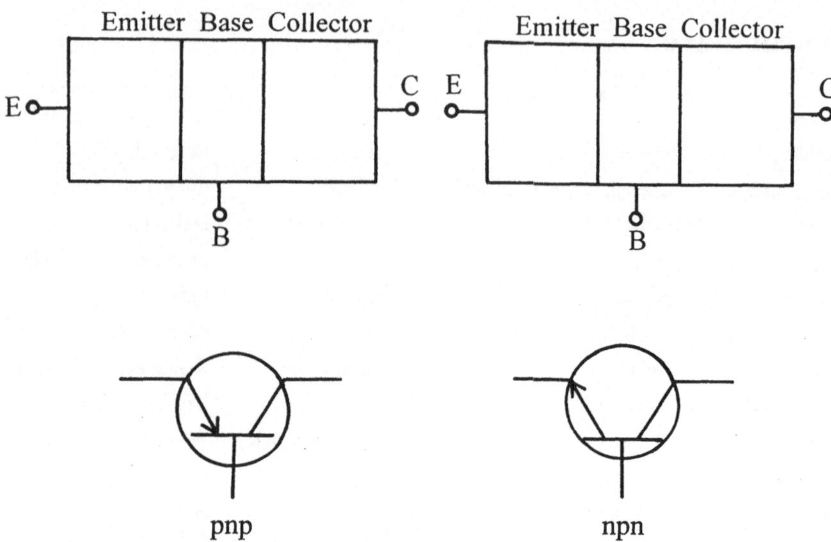

Fig 5.1 The two types of transistors and their symbols

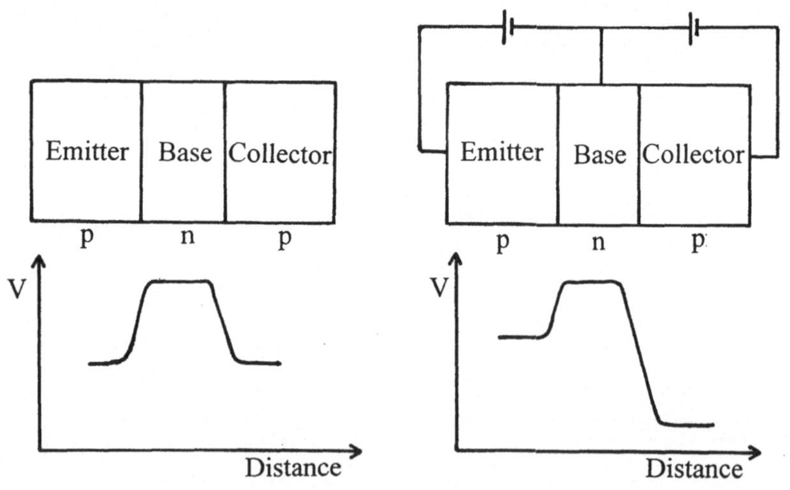

Fig 5.2 (a) Potential barriers in a transistor without bias

Fig 5.2 (b) Potential barriers in a transistor with bias

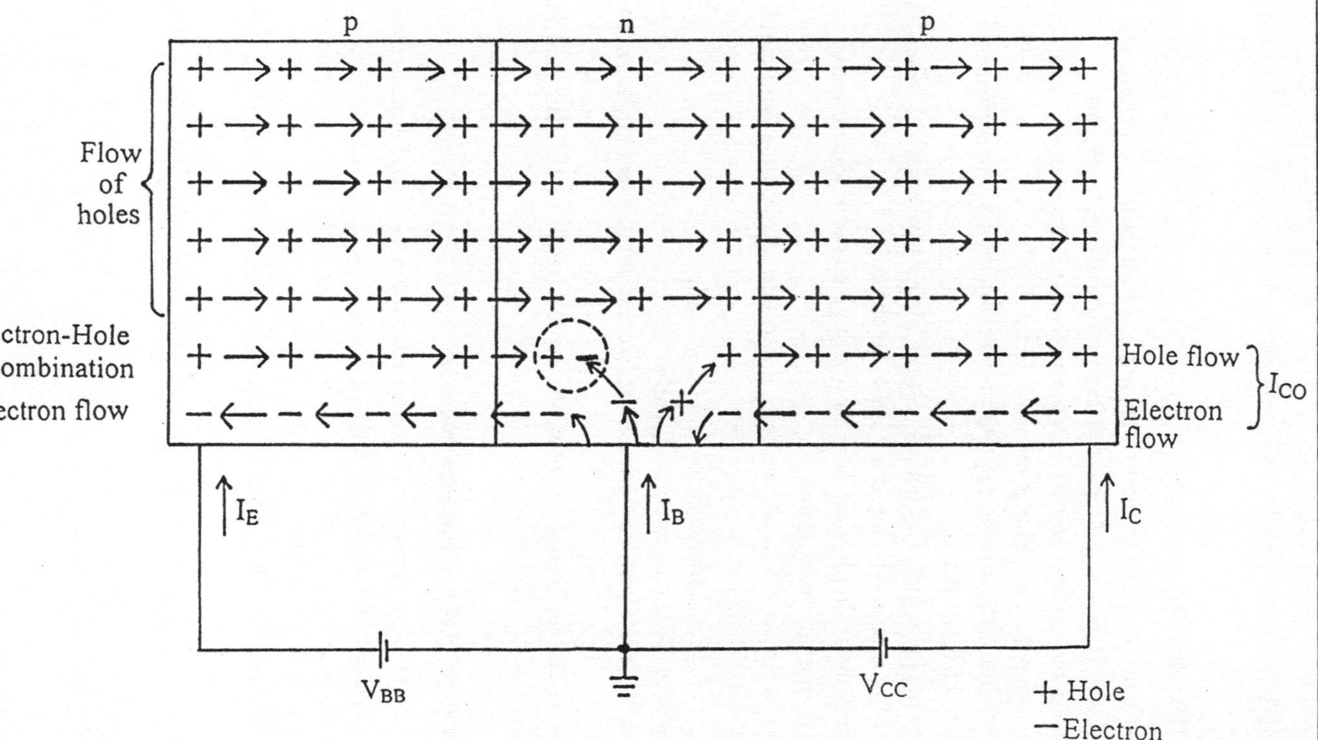

Fig 5.3 Physical processes involved in a transistor

Exercises V

1. Answer the following questions

a) What are the three semiconductor regions in a transistor called, and in what way is the central region different from the other regions ?
b) What is the difference between the potential barriers at the two junctions when the transistor is on open circuit ?
c) What kind of biasing is applied at each junction when a transistor is used as an amplifier ?
d) What happens to the potential barriers at the two junctions when the transistor is suitably biased for use as an amplifier ?
e) What is the main physical process which gives rise to the collector current I_C in a pnp transistor ?
f) What two main physical processes contribute towards the base current I_B ?
g) What is the relationship between I_C, I_B, and I_E ?
h) How does the leakage current I_{C0} arise ?
i) What is the relationship between the two current gain parameters α and β ?
j) What is the difference between the large signal current gain and the small signal current gain ?

2. Fill in the gaps in the following sentences:

a) The central region of a junction transistor has a ____ and a much lower ____ than the other two regions.
b) When a transistor is on ____ , there is no net flow of ____ across the junctions.
c) When a transistor is used as an amplifier, the first junction is ____ and the second junction is ____.
d) Holes in the ____ of a pnp transistor move across the first junction into the ____.
e) When holes reach the second junction ____ is in such a direction as to ____ them across the junction.
f) The flow of ____ out of the ____ constitutes part of the base current in a pnp transistor.

g) Most of the ____ in a pnp transistor flow from emitter to the collector without ____ in the base region.
h) The current I_{C0} is the ____ of the reverse biased _____ .
i) The ratio of the ____ current to the emitter current is represented by the ____ current gain parameter α.
j) The ratios of the currents are important ____ in studying ____ of a transistor.

3. Translate into English:

a) Falls keine äußeren Spannungen angelegt sind, überqueren keine Elektronen oder Löcher die beiden pn-Übergänge des Transistors. Bei einem symmetrischen Transistor sind die Potentialbarrieren an den beiden Übergängen gleich.
b) Wird ein Transistor als Verstärker eingesetzt, erhält der Emitter-Basis-Übergang eine Vorwärtsvorspannung und der Kollektor-Basis-Übergang eine Sperrvorspannung. Die Potentialbarriere am ersten Übergang wird gesenkt, während die Barriere am zweiten Übergang erhöht wird.
c) Löcher aus dem Emitter-Gebiet durchqueren den ersten Übergang in das Basis-Gebiet, da die Vorwärtsvorspannung die Potentialbarriere am Übergang unter den Gleichgewichtswert senkt.
d) Die meisten dieser Löcher driften durch das schmale Basis-Gebiet, ohne mit Elektronen in der Basis zu rekombinieren, da die Basis eine niedrige Elektronenkonzentration besitzt. Wenn die Löcher den zweiten Übergang erreichen, ist das Feld so gerichtet, daß sie durch den Übergang beschleunigt werden.
e) Transistoren werden so entworfen, daß der Basisstrom klein im Vergleich zum Emitterstrom ist. Dieses wird dadurch erreicht, daß das Basisgebiet schmal gemacht und mit einer geringen Störstellenkonzentration versehen wird. Die meisten Löcher fließen vom Emitter zum Kollektor, ohne mit Elektronen im Basisgebiet zu rekombinieren.

6 The basic amplifying action of a transistor

The transistor as an amplifier

The transistor is widely used as an amplifying device, and can provide voltage gain, current gain, and power gain. The basic amplifying action may be understood by considering the circuit of Fig 6.1. This circuit is known as the common base circuit due to the fact that the base terminal is common to both input and output circuits.

The input voltage is applied between the emitter and the base, while the output voltage is developed across the load resistor R_L. A small change in the input voltage ΔV_i produces a change in emitter current ΔI_E given by

$$\Delta I_E = \frac{\Delta V_i}{r_e}$$

where r_e is the input impedance of the circuit.

The change in emitter current produces a change in output voltage across the load resistor given by

$$\Delta V_o = -R_L \Delta I_C = -\alpha' R_L \Delta I_E$$

The voltage gain A_V, which is the ratio of the voltage across the load to the input voltage is given by

$$A_V = -\frac{\alpha' R_L \Delta I_E}{r_e \Delta I_E} = -\alpha' \frac{R_L}{r_e}$$

Since R_L can be made much larger than r_e, a high voltage gain is possible.

The common base circuit

There are three possible ways in which a transistor can be connected in an amplifier circuit. The circuits involved are called the common base, the common emitter, and the common collector circuits, depending on whether the base emitter or collector terminals of the transistor are common to both input and output circuits.

The common base circuit has a high voltage gain and a high power gain, but has a current gain of slightly less than unity. It has a low input impedance and a high output impedance. The circuit of a common base amplifier is shown in Fig 6.1.

The common collector or emitter follower circuit

A common collector circuit is shown in Fig 6.2. In this circuit the load resistor is connected between the emitter and the common or earth terminal. The voltage gain for this circuit is slightly less than unity but it has high values of current and power gain.

It has a high input impedance and a low output impedance. Such a circuit is often used as an impedance transformer for the matching of low impedance loads, like loudspeakers or transmission lines.

The common emitter circuit

The common emitter circuit shown in Fig 6.3 is the most frequently used of the three circuits mentioned. It has a high voltage gain, a high current gain, and a high power gain. The popularity of this circuit is due to the fact that its power gain is higher than for the other two circuits. The input and output impedances for this circuit are intermediate in value.

Vocabulary

across	über *pr*	**load resistor**	Lastwiderstand *m*
amplifier	Verstärker *m*	**match**	anpassen, übereinstimmen *v*
common base	Basisschaltung *f*	**mention**	erwähnen *v*
common collector	Kollektorschaltung *f*	**output impedance**	Ausgangsimpedanz *f*
common emitter	Emitterschaltung *f*	**popularity**	Beliebtheit *f*
common terminal	Massenanschlußpunkt *m*	**power gain**	Leistungsverstärkung *f*
current gain	Stromverstärkung *f*	**ratio**	Verhältnis *n*
emitter follower	Emitterfolger *m*	**slightly**	ein wenig, geringfügig *adv*
frequently	häufig *adv*	**transmission line**	Übertragungsleitung *f*
impedance transformer	Impedanzwandler *m*	**unity gain**	Verstärkung von Eins *f*
		value	Wert *m*
input impedance	Eingangsimpedanz *f*	**voltage gain**	Spannungsverstärkung *f*
involve	betreffen, enthalten *v*		

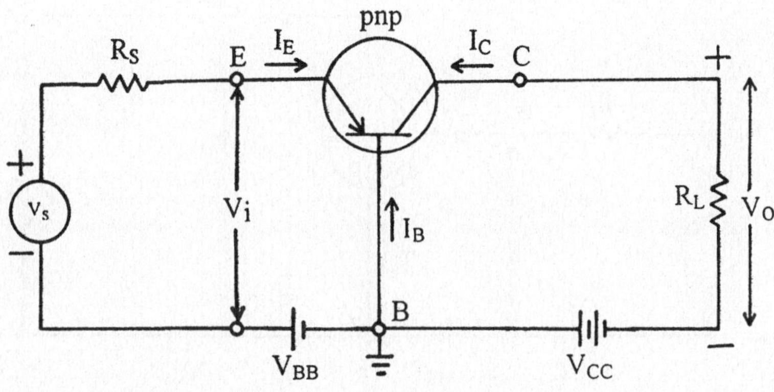

Fig 6.1 Basic amplifying action of a common base circuit using a pnp transistor

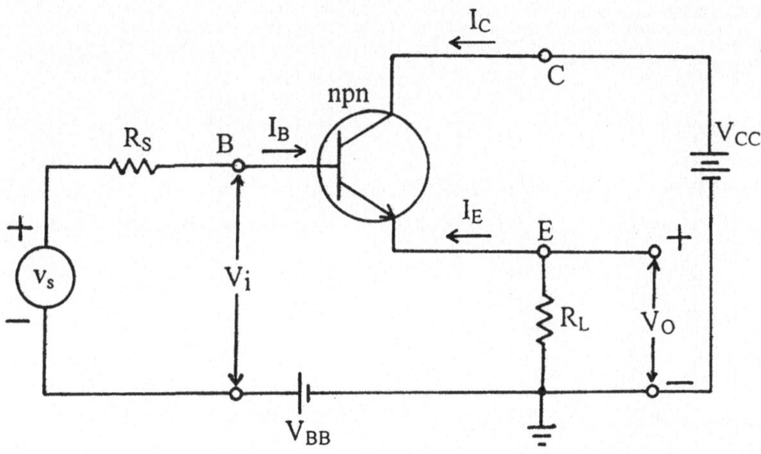

Fig 6.2 Basic common collector or emitter follower circuit using an npn transistor

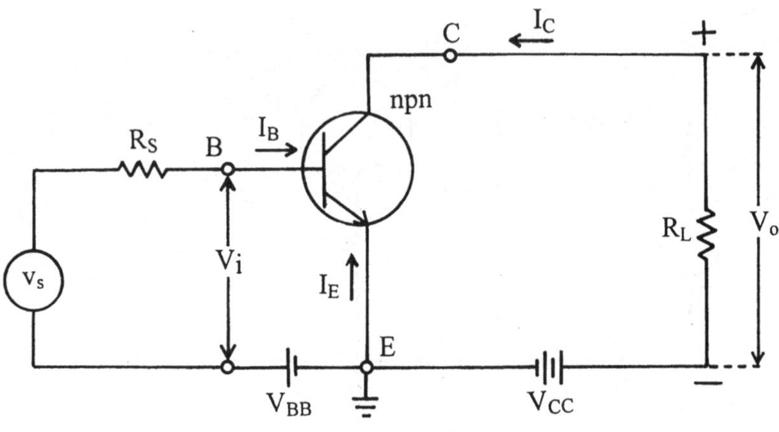

Fig 6.3 Basic common emitter circuit using an npn transistor

Exercises VI

1. Answer the following questions:

a) What types of gain can a transistor amplifier have ?
b) What are the essential features of a common base circuit ?
c) Where is the input voltage applied in a common base amplifier ?
d) Where is the output voltage developed ?
e) How is the voltage gain defined ?
f) Why does a common base amplifier have a high voltage gain ?
g) Where is the load connected in a common collector circuit ?
h) For what kind of application is a common collector amplifier particularly useful ?
i) What is the approximate voltage gain of the common collector circuit ?
j) Why is the common emitter circuit more popular than the other two circuits ?

2. Fill in the gaps in the following sentences:

a) The transistor is ____ used as an ____ .
b) In the ____ circuit, the base ____ is common to both input and output circuits.
c) In a common base circuit, the input voltage is ____ between the ____ and the base.
d) The output voltage is ____ across the ____ resistor.
e) The common base circuit has a ____ and a high power gain, but has a current gain of slightly less than ____.
f) In a common collector circuit, the ____ resistor is connected between the emitter and the ____.
g) A common ____ circuit is used as an ____ for matching low impedance loads.
h) A common base circuit has a low ____ and a high ____.
i) There are three possible ways in which ____ can be connected in an ____ circuit.
j) The popularity of the ____ circuit is due to the fact that its ____ is higher than for the other two circuits.

3. Translate into English:

a) Es gibt drei Möglichkeiten, einen Transistor in einer Verstärkerschaltung zu beschalten. Die Schaltungen heißen Basis-, Emitter- und Kollektorschaltung, abhängig davon, ob der Basis-, Emitter- oder Kollektoranschluß des Transistors dem Eingangs- und dem Ausgangskreis gemeinsam ist.

b) Bei der Kollektorschaltung, oder Emitterfolger, wird der Lastwiderstand zwischen Emitter und Masse angeschlossen. Die Spannungsverstärkung dieser Schaltung ist etwas kleiner als Eins, aber sie besitzt eine hohe Strom- und Leistungsverstärkung.

c) Die Kollektorschaltung hat eine hohe Eingangsimpedanz und eine geringe Ausgangsimpedanz. Eine solche Schaltung wird häufig als Impedanzwandler zur Anpassung von niederimpedanten Lasten eingesetzt.

d) Die Eingangsspannung wird zwischen Emitter und Basis angelegt, während sich die Ausgangsspannung am Lastwiderstand einstellt. Eine kleine Änderung der Eingangsspannung verursacht eine große Änderung des Emitterstroms.

e) Die Emitterschaltung wird am häufigsten von allen drei Schaltungen eingesetzt. Die Beliebtheit dieser Schaltung beruht auf der Tatsache, daß ihre Leistungsverstärkung höher ist als die der anderen beiden Schaltungen.

7 Transistor characteristics and the operating point

Transistor characteristics for the common emitter circuit

When a transistor is used in a circuit, the D.C operating conditions have to be first optimized so that the transistor will operate as linearly as possible. It is desirable that the largest possible output voltage be produced with the minimum of distortion. The optimization of the D.C operating conditions is done by a graphical method which uses two families of characteristic curves called the input and output characteristics.

The common emitter input characteristics

These consist of a family of curves showing how I_B varies when V_{BE} is changed, as shown in Fig 7.1. Each curve is drawn for a fixed value of V_{CE}. The curve for $V_{CE} = 0$ (collector short circuited to emitter) corresponds to a forward biased diode. The other curves are only slightly different.

The common emitter output characteristics

These curves show how I_C varies when V_{CE} is changed. Each curve is drawn for a different value of I_B as shown in Fig 7.2. It is convenient to divide the area covered by the curves into three regions, i.e. saturation region, cut-off region and active region. In the saturation region, both junctions are forward biased. This region is shown shaded in Fig 7.2.

The cut-off region is defined as the region for which $I_E = 0$. For practical purposes this can be considered to be the region below the curve $I_B = 0$. In the active region, the emitter junction is forward biased and the collector junction reverse biased. The dotted hyperbola drawn on the output characteristics is the locus corresponding to the maximum power rating of the transistor given by $P_C = V_{CE}\ I_C$. The operating region of the transistor must be kept below this curve to avoid damaging the transistor.

To obtain maximum linearity of operation, the transistor must be kept within the active region and below the maximum power rating curve. Other limitations are that the maximum allowed values of I_C and V_{CE} must not be exceeded. These limiting values are usually provided by the manufacturer.

The load line and the operating point

It is convenient and normal to use a single D.C voltage supply instead of two batteries to obtain the biasing voltages for a transistor circuit. The simplest type of circuit using a single voltage supply is the fixed bias circuit shown in Fig 7.3. In designing such a circuit, the D.C operating conditions have to be first chosen, and this involves the following steps:

1. **Choosing a suitable D.C supply voltage.** A suitable supply voltage can be chosen using the transistor data supplied by the manufacturer.

2. **Choosing a suitable load resistor by drawing a load line.** To find a suitable value of load resistor, a straight line called the load line is drawn on the output characteristics as shown in Fig 7.2. The equation of the load line is

$$V_{CE} = V_{CC} - R_L I_C$$

   It will be seen that $V_{CE} = V_{CC}$ when $I_C = 0$. From this it follows that all load lines must pass through the point corresponding to V_{CC} which is point B on the diagram. Two such load lines are shown in the figure. The load line cuts the I axis at $V_{CE} = 0$. This is when

$$I_C = \frac{V_{CC}}{R_L}$$

   It is clear from the above expression that once the load line has been drawn, the value of the corresponding load resistor can be found. The load line BC is tangential to the maximum power hyperbola, and all other load lines must lie to the left of this. A good choice of load line would be BD. The value of R_L corresponding to this line is 170 ohm.

3. **Choice of operating point.** The point O would be a good choice of operating point for the load line BD. This corresponds to a base current of 100 microamps. When the bias current increases or decreases by 60 microamps, we obtain the points E and F which correspond to approximately equal changes in collector current.

4. **Choice of bias resistor.** The value of the bias resistor R_B corresponding to a base bias current of 100 microamp (point 0) is obtained as follows:

$$R_B = \frac{V_{CC}}{I_B} = \frac{8V}{100\mu A}$$

$$R_B \approx 80k\Omega$$

Bias stability

The fixed bias circuit shown in Fig 7.3 has poor stability. Changes in temperature cause changes in the current I_{C0}. Also the variation in the values of β for different transistors of the same type can be large. Consequently the operating point changes if the temperature changes, or if the transistor is replaced. A better biasing circuit called self bias or emitter bias which partially compensates for these changes is shown in Fig 7.4.

Vocabulary

approximately	ungefähr *adv*	**linearly**	linear *adv*
choice	Wahl *f*	**limitation**	Begrenzung *f*
circuit	Schaltung *f*	**load line**	Belastungskennlinie *f*
clear	klar *adj*	**operating condition**	Betriebsbedingung *f*
compensate	ausgleichen *v*	**operating point**	Arbeitspunkt *m*
consider	bedenken *v*	**optimize**	optimieren *v*
correspond	entsprechen *v*	**output characteristic**	Ausgangskennlinie *f*
curve	Kurve *f*	**region**	Bereich *m*, Gebiet *n*
cut-off	abschalten, abschneiden *v*	**replace**	ersetzen *v*
damage	Schaden *m*	**saturation region**	Sättigungsbereich *m*
define	definieren *v*	**short circuit**	Kurzschluß *m*
distortion	Verzerrung *f*	**stability**	Stabilität *f*
dotted	gestrichelt *adj*	**suitable**	geeignet, passend *adj*
exceed	überschreiten *v*	**supply voltage**	Betriebsspannung *f*
input characteristic	Eingangskennlinie *f*	**tangential**	tangential *adj*

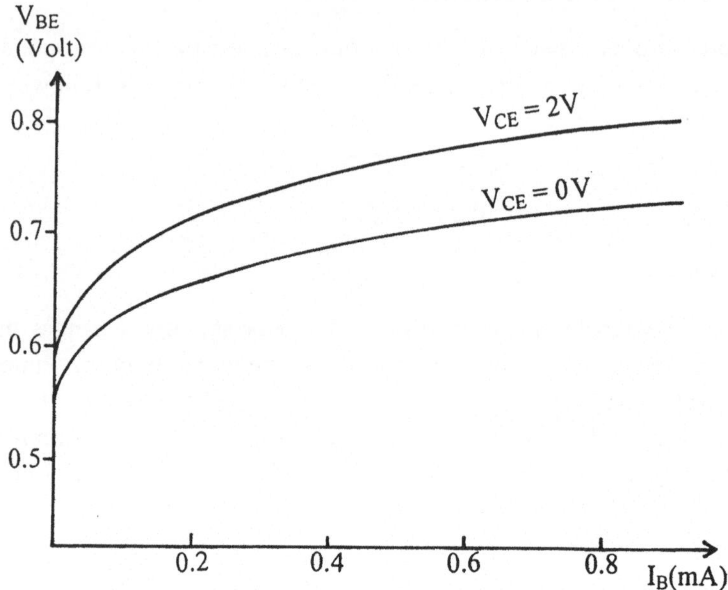

Fig 7.1 Common emitter input characteristics for a npn transistor

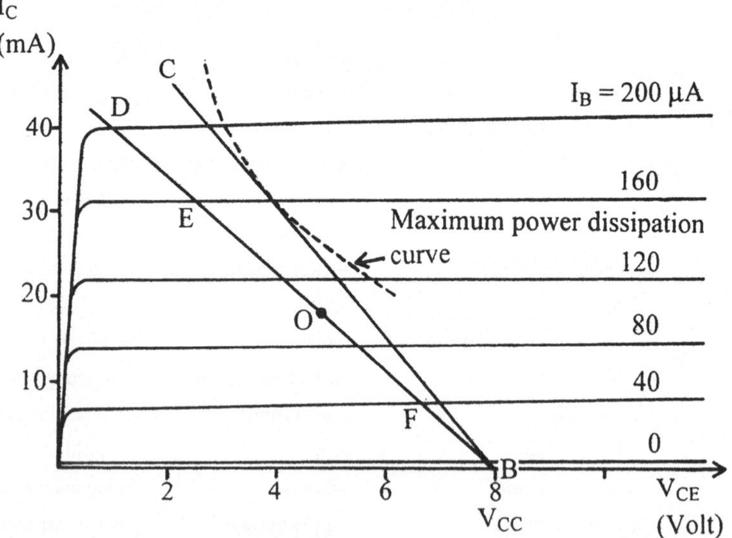

Fig 7.2 Common emitter output characteristics for a npn transistor

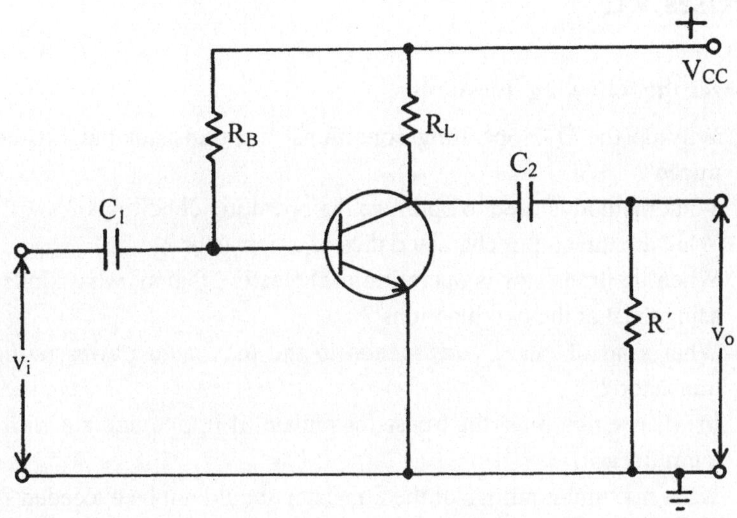

Fig 7.3 A fixed bias circuit

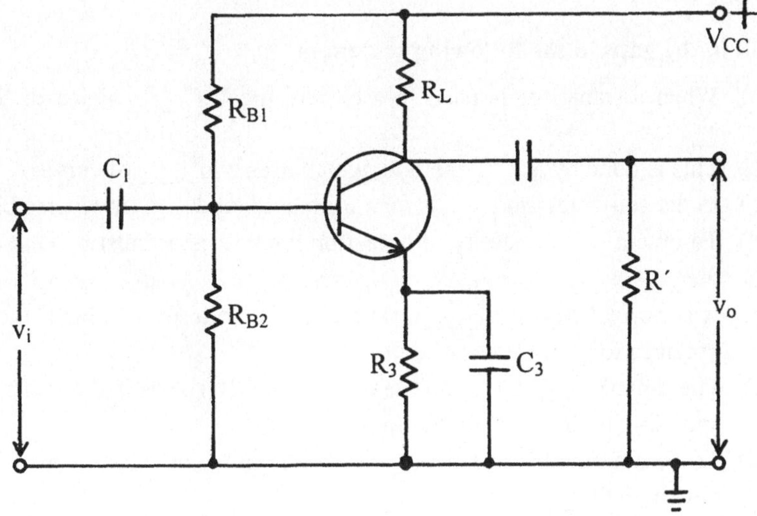

Fig 7.4 A self bias circuit

Exercises VII

1. Answer the following questions:

a) Why do the D.C operating conditions in a transistor have to be optimized ?
b) What method is used to optimize the operating conditions ?
c) What are the output characteristics of a transistor ?
d) When the transistor is operating in the active region, what kinds of biasing exist at the two junctions ?
e) What kind of curve corresponds to the maximum power rating of a transistor ?
f) In what region must the transistor remain, if it is to operate with maximum linearity ?
g) What maximum ratings of the transistor should not be exceeded ?
h) What steps are involved in determining the D.C operating conditions ?
i) How is the value of a suitable load resistor found ?
j) Why is the self bias circuit better than the fixed bias circuit ?

2. Fill in the gaps in the following sentences

a) When a transistor is used in a circuit, the D.C ____ have to be first ____.
b) This is done by a ____ method, which uses two ____ of curves.
c) In the active region ____ is forward biased, and ____ is reverse biased.
d) To obtain ____ linearity of operation the transistor must be kept within the ____.
e) It is normal to use a ____ instead of two batteries to obtain the ____ voltages for a transistor circuit.
f) The dotted ____ drawn on the output characteristics is the locus corresponding to the ____ of the transistor.
g) To find a suitable value of ____, a straight line called a load line is drawn on the ____.
h) The ____ characteristics show how I_C varies when ____ is changed.
i) Other limitations are that the maximum ____ of I_C and V_{CE} must not be ____.
j) If the ____ changes or if the transistor is changed, the ____ changes.

3. Translate into English:

a) Wird ein Transistor in einer Schaltung eingesetzt, müssen zunächst die Gleichstrombetriebsbedingungen optimiert werden, so daß der Transistor möglichst linear betrieben wird. Die höchste mögliche Ausgangsspannung sollte mit minimaler Verzerrung erzeugt werden.

b) Um ein Maximum an Linearität im Betrieb zu erreichen, muß der Transistor im aktiven Bereich unterhalb der maximalen Leistungskurve betrieben werden.

c) Normalerweise wird eine einzelne Gleichstromquelle statt zweier Batterien eingesetzt, um die Vorspannungen für eine Transistorschaltung einzustellen. Die einfachste Variante einer solchen Schaltung ist die Schaltung mit fester Vorspannung in Abb. 7.3.

d) Eine geeignete Versorgungsspannung kann anhand des vom Hersteller gelieferten Transistor-Datenblatts gewählt werden. Um einen geeigneten Wert für den Lastwiderstand zu bestimmen, wird durch das Ausgangskennlinienfeld eine Gerade gelegt.

e) Die Schaltung mit fester Vorspannung hat eine geringe Stabilität. Temperaturänderungen verursachen eine Änderung des Stroms I_{C0}. Auch könnnen die Abweichungen der β-Werte bei verschiedenen Transistoren des gleichen Typs groß sein.

8 Transistor equivalent circuits

The first step in designing a transistor amplifier stage is to establish the D.C operating conditions by using a graphical method as shown in the last chapter. After this has been done, small signals which are applied to the input of a transistor circuit will be amplified with reasonable linearity.

The quantitative analysis of a transistor circuit can now be carried out analytically by using equivalent circuits which represent the behaviour of the transistor in the active region. The small signal parameters used in the equivalent circuits are usually supplied by the manufacturer. They can also be determined experimentally by the user. The most commonly used equivalent circuit is the hybrid equivalent circuit.

The hybrid circuit and the h-parameters

If a transistor circuit is replaced by an equivalent four terminal network as shown in Fig 8.1, the behaviour of the network can be studied in terms of two voltages and two currents, (i.e. v_1, v_2 and i_1, i_2). If the two variables i_1 and v_2 are selected as independent variables, we can write

$$v_1 = h_{11}i_1 + h_{12}v_2$$
$$i_2 = h_{21}i_1 + h_{22}v_2$$

If the circuit is assumed to be linear, then the quantities h_{11}, h_{12}, h_{21}, h_{22} are constants and are called hybrid parameters, because they are dimensionally different from each other. From the above equations the hybrid parameters may be defined as follows

$$h_{11} = \left.\frac{v_1}{i_1}\right|_{v_2=0}$$ input resistance with output short-circuited (ohm)

$$h_{12} = \left.\frac{v_1}{v_2}\right|_{i_1=0}$$ reverse open circuit voltage amplification (dimensionless)

$$h_{21} = \left.\frac{i_2}{i_1}\right|_{v_2=0}$$ short circuit current gain (dimensionless)

$$h_{22} = \left.\frac{i_2}{v_2}\right|_{v_1=0}$$ output conductance with input short-circuited (ohm^{-1}, mho)

For transistor circuits the subscripts e, b, c are added to indicate the kind of circuit, common emitter, common base, or common collector. In addition, the following notation is used for the subscripts:

i = 11 = input, o = 22 = output

f = 21 = forward transfer, r = 12 = reverse transfer

If we represent small changes in voltage and current about the operating point by v_b, v_c, i_b, and i_c, then we can write the following equations for a common emitter circuit

$$v_b = h_{ie} i_b + h_{re} v_c$$
$$i_c = h_{fe} i_b + h_{oe} v_c$$

The hybrid model for the common emitter circuit is shown in Fig 8.2. Since h_{re} is usually very small this parameter can in most cases be omitted and the simplified model shown in Fig 8.3 can be used.

Analysis of a transistor amplifier circuit

The use of h-parameters in the analysis of a simple transistor amplifier circuit is shown below. The circuit used is the one given in Fig 8.2.

The current gain A_I

The current gain A_I is given by

$$A_I = \frac{i_L}{i_1} = -\frac{i_2}{i_1}$$

From Fig 8.2, we can write

$$i_2 = h_f i_1 + h_o v_2$$

Since

$$v_2 = -i_2 Z_L$$

$$A_I = \frac{i_L}{i_1} = -\frac{i_2}{i_1} = -\frac{h_f}{1 + h_0 Z_L}$$

The voltage gain A_V

The voltage gain A_V is given by

$$A_V = \frac{v_2}{v_1}$$

Since $$v_2 = -i_2 Z_L$$

and $$A_I = -\frac{i_2}{i_1}$$

it follows that $$v_2 = A_I i_1 Z_L$$

Since $$v_1 = i_1 Z_i$$

it follows that $$A_V = A_I \frac{Z_L}{Z_i}$$

The input impedance Z_i

The input impedance Z_i is given by

$$Z_i = -\frac{v_1}{i_1}$$

Since

$$v_1 = h_i i_1 + h_r v_2$$

it follows that

$$Z_i = h_i + \frac{h_r v_2}{i_1}$$

Also

$$v_2 = -i_2 Z_L = A_I i_1 Z_L$$
$$Z_i = h_i + h_r A_I Z_L$$

Substituting

$$A_I = -\frac{h_f}{1 + h_0 Z_L}$$

we have

$$Z_i = h_i - \frac{h_f h_r}{h_0 + 1/Z_L}$$

Vocabulary

amplifier stage	Verstärkerstufe *f*	**independent**	unabhängig *adj*
analysis	Analyse, Zerlegung *f*	**indicate**	anzeigen *v*
behaviour	Verhalten *n*, Benehmen *n*	**linearity**	Linearität *f*
define	definieren *v*	**manufacturer**	Hersteller *m*
design	entwerfen *v*	**notation**	Darstellung *f*
determine	bestimmen *v*	**omit**	weglassen, übergehen *v*
dimension	Dimension *f*, Maß *n*	**reasonable**	vernünftig *adj*
equivalent circuit	Ersatzschaltung *f*	**represent**	vertreten, darstellen *v*
establish	errichten *v*, etablieren *v*	**small signal parameter**	Kleinsignal-Parameter *m*
four terminal network	Vierpol *m*	**step**	Schritt *m*
hybrid parameter	Hybridparameter *m*	**subscript**	Index *m*
hybrid equivalent circuit	Hybrid-Ersatzschaltbild *f*	**variable**	Variable *f*

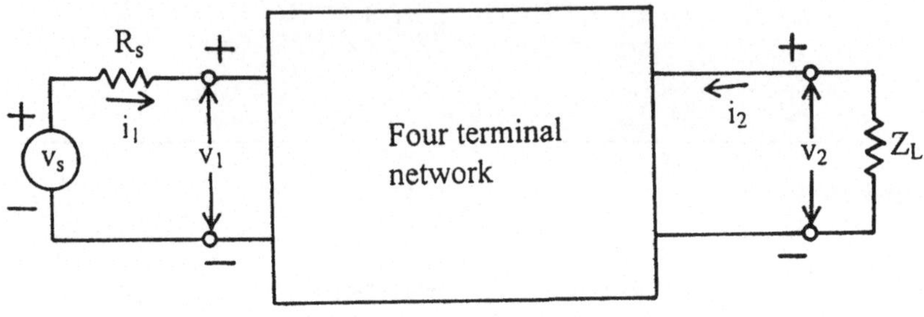

Fig 8.1 Four terminal network used to replace a transistor circuit

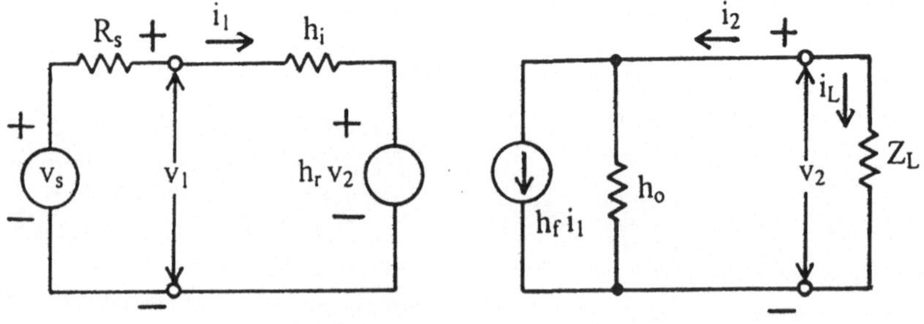

Fig 8.2 The hybrid equivalent circuit

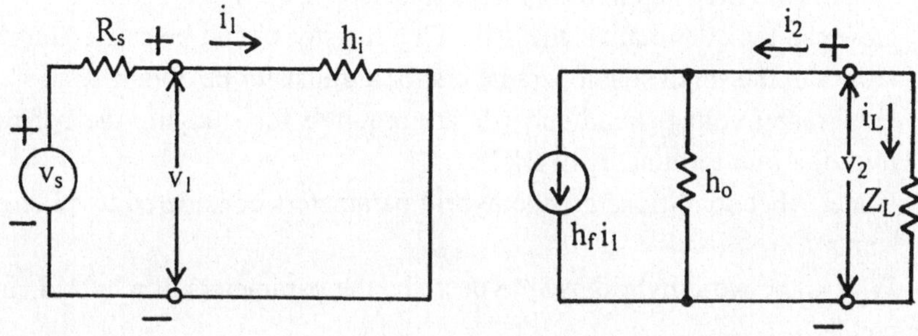

Fig 8.3 Simplified hybrid equivalent circuit

Exercises VIII

1.Answer the following questions:

a) What is the first step in designing a transistor amplifier stage ?
b) How can the quantitative analysis of a transistor circuit be carried out ?
c) How can the small signal parameters for a transistor be obtained ?
d) How many voltages and currents are required for studying the behaviour of a four terminal network ?
e) Under what conditions are the hybrid parameters considered to be constants ?
f) Why is the word hybrid used to describe the parameters of a hybrid circuit ?
g) What do the subscripts e,b,c, indicate when used with the hybrid parameters ?
h) Define the parameters h_{21} and h_{22}.
i) What do the letters i, f, o, r, mean when used with the hybrid parameters?
j) Which hybrid parameter can be omitted in the simplified hybrid model?

2. Fill in the gaps in the following sentences :

a) The first step in ____ a transistor amplifier stage is to establish the D.C ____ by using a graphical method.
b) When this has been done, ____ applied to the input of a transistor circuit are amplified with reasonable ____.
c) The analysis of a transistor circuit can be carried out by using ____ which ____ the behaviour of the transistor.
d) The small signal ____ used in the equivalent circuits are usually supplied by ____.
e) The most commonly used equivalent circuit is the ____ , and its parameters can be ____ by experiment.
f) The transistor circuit is ____ by an equivalent ____.
g) If the circuit is ____, then the ____ are constants.
h) The parameter h_{11} is defined as the ____ with ____ short-circuited.
i) For transistor circuits the ____ e, b, c, are added to ____ the kind of circuit.

j) In the simplified hybrid equivalent circuit, the ____ can be omitted because it is ____.

3. Translate into English:

a) Der erste Schritt beim Entwurf einer Verstärkerstufe besteht darin, die Gleichstromarbeitsbedingungen graphisch zu bestimmen. Ist dieses erledigt, so werden an den Eingang angelegte Kleinsignale mit annehmbarer Linearität verstärkt.
b) Eine Transistorschaltung kann durch ein äquivalentes Vierpolnetzwerk ersetzt werden. Das Verhalten eines solchen Netzwerkes kann anhand zweier Spannungen und zweier Ströme untersucht werden.
c) Die im Ersatzschaltbild verwendeten Kleinsignalparameter werden üblicherweise vom Hersteller angegeben. Sie können auch vom Anwender experimentell für einen vorliegenden Transistor ermittelt werden.
d) Um eine maximale Linearität des Betriebes zu erhalten, muß der Transistor im Betriebsbereich unterhalb der Kurve der maximalen Verlustleistung gehalten werden. Die maximal zulässigen Werte von I_C und V_{CE}, wie vom Hersteller angegeben, sollten nicht überschritten werden.
e) Der Hybridparameter h_{11} ist als Eingangswiderstand bei ausgangsseitigem Kurzschluß definiert, während der Parameter h_{21} das Verhältnis von Ausgangs- zu Eingangsstrom angibt und als Kurzschlußstromverstärkung bezeichnet wird.

9 The frequency response of a transistor amplifier

Wideband amplifiers

Wideband amplifiers can amplify over a wide band of frequencies from a few hertz to several megahertz. A typical example of such an amplifier is the video amplifier in a television receiver. A nonsinusoidal input signal is composed of many frequencies, and if the output is to be an exact replica of the input, then the following conditions must be satisfied.

The gain of the amplifier should be independent of the frequency, and the phase shift should be proportional to the frequency. This means that the time delay should be the same for all frequency components of a signal.

Frequency response characteristics of an amplifier

The gain and phase shift produced by an amplifier, usually vary with the frequency of the input signal. The gain A may be considered to be a complex number, whose magnitude and phase angle are functions of frequency. Curves showing the variation of gain and phase with frequency are called the amplitude-frequency and the phase-frequency characteristics of the amplifier. The circuit of a typical resistance-capacity (RC) coupled transistor amplifier is shown in Fig 9.1. The amplitude-frequency characteristic of such an amplifier is shown in Fig 9.2. The characteristic can be divided into three regions, midfrequency, low frequency, and high frequency.

Midfrequency region

The midfrequency region usually extends over several decades, and in this region, the gain and time delay remain reasonably constant.

Low frequency region

In the low frequency region, the gain decreases with decreasing frequency, due to the effect of the external coupling capacitor C_1, and the emitter resistor bypass capacitor C_2.The amplifier behaves like the simple highpass circuit shown in Fig 9.3.

For the circuit shown

$$V_0 = \frac{R_1}{R_1 + 1/j\omega C_1} V_i$$

If $A_L = V_0/V_i$ is the low frequency voltage gain, then

$$A_L = \frac{1}{1 - j(f_L/f)}$$

where
$$f_L = \frac{1}{2\pi C_1 R_1}$$

The magnitude and phase of A_L are given by

$$|A_L| = \frac{1}{\sqrt{1 + (f_L/f)^2}}$$

and
$$\Theta_L = \arctan\frac{f_L}{f}$$

When $f = f_L$ and $A_L = 1/\sqrt{2}$ times the midfrequency gain, the drop in gain corresponds to a reduction in decibels of 3 dB. Therefore f_L is called the low 3 dB frequency.

High frequency region

The response of an amplifier in the high frequency region is similar to that of a low pass filter as shown in Fig 9.4.

$$V_0 = \frac{1/j\omega C_2}{R_2 + 1/j\omega C_2} V_i = \frac{1}{1 + j\omega R_2 C_2} V_i$$

The magnitude A_H and the phase angle Θ_H of the gain are

$$|A_H| = \frac{1}{\sqrt{1 + (f/f_H)^2}}$$

and
$$\Theta_H = -\arctan\frac{f}{f_H}$$

where $$f_H = \frac{1}{2\pi R_2 C_2}$$

The frequency f_H is called the upper 3 dB frequency, because the gain at this frequency is $1/\sqrt{2}$ times the midfrequency gain.

Bandwidth

The frequency range between f_L and f_H is called the bandwidth of the amplifier. Frequencies that lie within this range will be amplified without excessive distortion.

The step response of an amplifier

Another way of assessing the behaviour of an amplifier, is to consider the response of the amplifier to a step waveform. The high frequency response is closely related to the leading edge of the output waveform and the low frequency response is closely related to the sag in the flat portion of the output waveform.

Rise time

The behaviour of the amplifier at high frequencies has been seen to be similar to that of the low pass circuit of Fig 9.4. When a step waveform is applied to such a circuit, the capacitor charges exponentially with a time constant CR as shown in Fig 9.5. The output is given by

$$V_0 = V\left(1 - e^{-t/R_2C_2}\right)$$

The time t_r taken for the voltage to rise from one-tenth to nine-tenths of its final value is known as the rise time. It can be shown that

$$t_r = 2.2R_2C_2 = \frac{0.35}{f_H}$$

The rise time is a measure of how fast an amplifier can respond to a sudden change in input voltage. It can be seen that t_r is inversely proportional to f_H.

Sag or droop of a waveform

The behaviour of the amplifier at low frequencies is similar to the response of the highpass circuit of Fig 9.3. If a step voltage is applied to the highpass circuit

$$V_0 = Ve^{-t/R_1C_1}$$

For times t_1 which are small compared to the time constant R_1C_1,

$$V_0 = V\left(1 - {t_1}/{R_1C_1}\right)$$

This is shown in Fig 9.6, in which a sag in the waveform with increasing time may be observed.

The percentage sag in time t_1 is

$$P = \frac{V - V'}{V} \times 100\% = \frac{t_1}{R_1C_1} \times 100\%$$

If a square wave is used for testing purposes $t_1 = T/2$ and $f = 1/T$ where f is the frequency of the square wave

$$P = \frac{T}{2R_1C_1} = \frac{\pi f_L}{f} \times 100\%$$

The sag of the waveform is therefore directly proportional to the lower 3 dB frequency.

Vocabulary

assess	einschätzen, bewerten *v*	**frequency component**	Frequenzkomponente *f*
bandwidth	Bandbreite *f*	**frequency response**	Frequenzgang *m*
behave	sich verhalten *v*	**highpass circuit**	Hochpaß Schaltung *f*
bypass capacitor	Ableitkondensator *m*	**phase shift**	Phasenverschiebung *f*
complex numbers	komplexe Zahlen *n*	**replica**	Kopie, Nachbildung *f*
condition	Bedingung, Voraussetzung *f*	**rise time**	Anstiegszeit *f*
correspond	entsprechen v	**satisfy**	genügen, befriedigen *v*
coupling capacitor	Koppelkondensator *m*	**sag**	Senkung *f*, Durchhang *m*
decade	Dekade *f*	**step response**	Sprungantwort *f*
decrease	verkleinern *v*	**time delay**	Zeitverzögerung *f*
droop	Abfall *m*, Abweichung *f*	**video amplifier**	Videoverstärker *m*
exact	genau adj	**wavefunction**	Wellenfunktion *f*
example	Beispiel *n*	**wideband amplifier**	Breitbandverstärker *m*
extend	ausdehnen *v*		

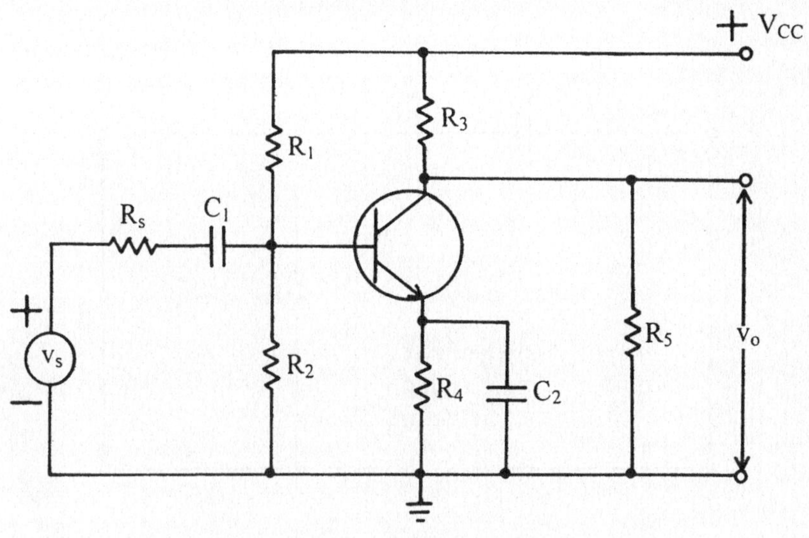

Fig 9.1 Typical RC coupled amplifier circuit

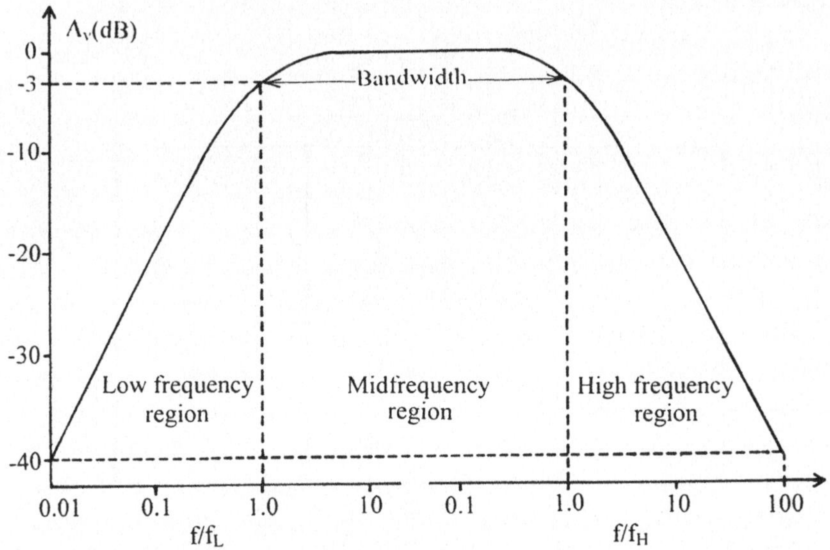

Fig 9.2 Amplitude-frequency (Bode) characteristic of an RC amplifier plotted on a log-log scale

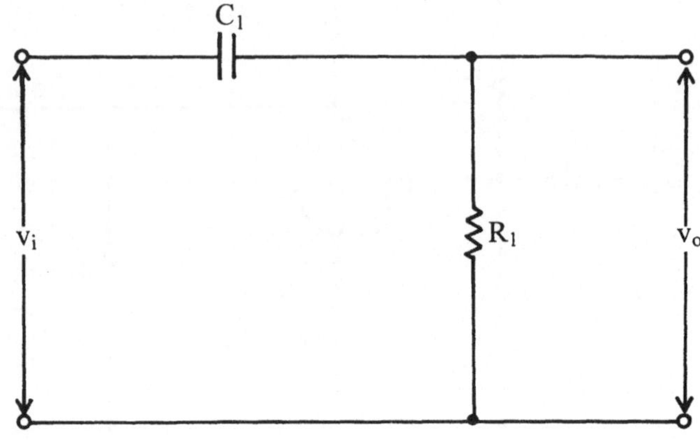

Fig 9.3 High pass circuit used to calculate the low frequency response of a RC amplifier

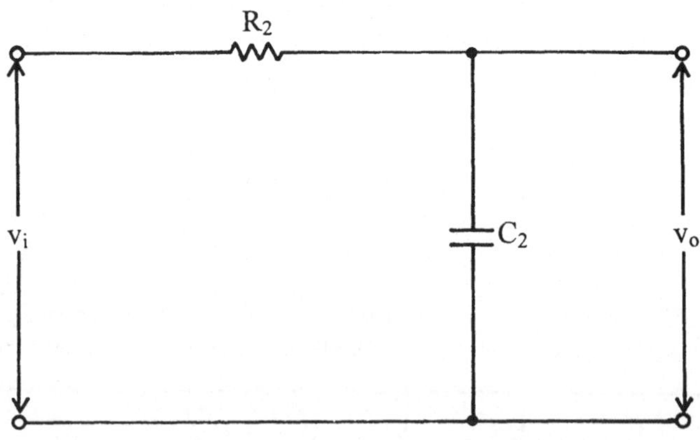

Fig 9.4 Low pass circuit used to calculate the high frequency response of a RC coupled amplifier

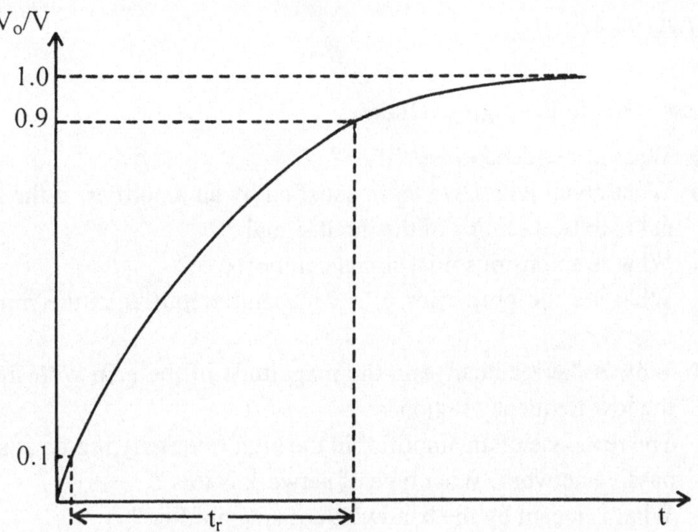

Fig 9.5 Response of a low pass RC circuit to a step voltage

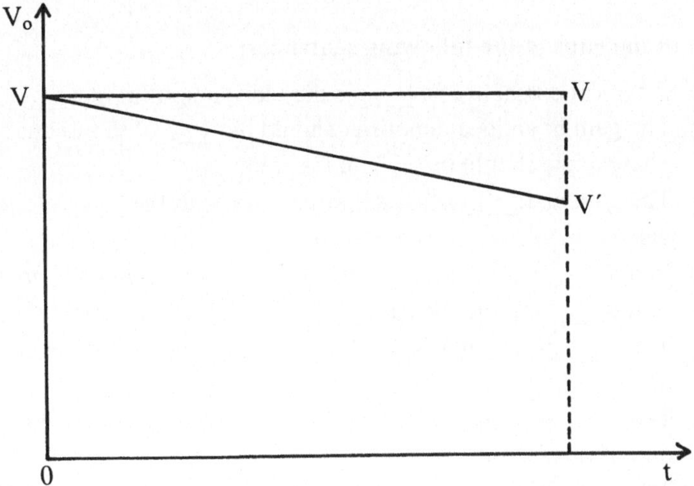

Fig 9.6 Sag in the output of an RC amplifier when a step voltage is applied to the input

Exercises IX

1. Answer the following questions:

a) What is a wideband amplifier ?
b) What conditions have to be satisfied by an amplifier, if the output signal is to be a replica of the input signal ?
c) What is a nonsinusoidal signal composed of ?
d) What are the properties of a wideband amplifier in the midfrequency region ?
e) Why is there a change in the magnitude of the gain with frequency in the low frequency region ?
f) The response of an amplifier in the high frequency region is similar to a passive network. What type of network is this ?
g) What is meant by the bandwidth of an amplifier ?
h) What is meant by the rise time of an amplifier ?
i) What property of an amplifier determines the sag in the output which results when a step voltage is applied to its input ?
j) What are the 3 dB frequencies of an amplifier ?

2. Fill in the gaps in the following sentences:

a) A typical example of a ____ is the video amplifier in a ____.
b) The gain of an ideal amplifier should be ____ of the frequency, and the phase angle should be ____ to the frequency.
c) The gain and ____ of an amplifier vary with the ____ of the input signal.
d) In the ____ region, the gain and ____ remain reasonably constant.
e) In the ____ region, the gain ____ with decreasing frequency.
f) The ____ of an amplifier in the high frequency region is similar to that of a ____.
g) The ____ response of an amplifier is closely related to the ____ of the output waveform.
h) The ____ of an amplifier is the time taken for the voltage to rise from ____ of its final value.
i) The rise time is a measure of how ____ an amplifier can ____ to a sudden change in input voltage.

j) Another way of assessing the ____ of an amplifier is to consider its response to a ____.

3. Translate into English:

a) Breitbandverstärker können einen breiten Frequenzbereich von wenigen Hertz bis einigen Megahertz verstärken. Ein typisches Beispiel eines solchen Verstärkers ist der Videoverstärker eines Fernsehempfängers.
b) Die Verstärkung sollte von der Frequenz unabhängig und die Phasenverschiebung der Frequenz proportional sein, d.h., daß die Verzögerungszeit für alle Frequenzkomponenten des Signals gleich sein sollte.
c) Im Niederfrequenzbereich nimmt die Verstärkung mit der Frequenz ab, aufgrund der Wirkung des Koppelkondensators sowie des Ableitkondensators am Emitterwiderstand.
d) Die Schaltung eines typischen RC-gekoppelten Verstärkers ist in Abb. 9.1 dargestellt. Die Amplitudenfrequenzgang-Kennlinie kann in drei Bereiche aufgeteilt werden, den niederfrequenten Bereich, den hochfrequenten Bereich und den mittleren Frequenzbereich.
e) Die von der Ausgangsspannung benötigte Zeit, um von einem Zehntel auf neun Zehntel des Endwertes zu steigen, ist als Anstiegszeit bekannt. Die Anstiegszeit ist ein Maß dafür, wie schnell ein Verstärker auf eine plötzliche Änderung der Eingangsspannung reagieren kann.

10 The transistor at high frequencies

The currents flowing through a transistor arise from the flow of charge carriers between the different regions of the transistor.This process takes time and the response of the transistor to changes in input voltages and currents is not instantaneous. The result is that there is a gradual decrease of the short-circuit current gain parameter at high frequencies. The equivalent circuit used to represent the high frequency behaviour of a transistor has therefore to be different from the circuit used at low frequencies.

The hybrid-π circuit

An equivalent circuit that has been widely used to represent the behaviour of a transistor at high frequencies is the hybrid-π circuit shown in Fig 10.1. The resistances and capacitances used in the circuit are assumed to be independent of frequency. The values of the resistances and conductances can be obtained from the low frequency h-parameters. If the low frequency h-parameters for the common emitter circuit are known for a given collector current I_C, then the resistances and conductances that are used in the hybrid-circuit can be calculated from the following equations

$$g_m = \frac{I_C}{V_T} \quad \text{where} \quad V_T = \frac{kT}{e} = \frac{T}{11600}$$

$$r_{b'e} = \frac{h_{fe}}{g_m}$$

$$r_{bb'} = h_{ie} - r_{b'e}$$

$$r_{b'c} = \frac{r_{b'e}}{h_{re}}$$

$$g_{ce} = h_{oe} - (1 + h_{fe})g_{b'c} = \frac{1}{r_{ce}}$$

The hybrid-π circuit capacitances

There are two capacitances in the hybrid-π circuit. The capacitance C_C is the measured value of the collector to base output capacitance with open input (i.e. $I_E = 0$). This is usually given in the manufacturer's data sheets as C_{ob}.

The capacitance C_e is the sum of the emitter diffusion capacitance C_{De} and the emitter junction capacitance C_{Te} where C_{De} is usually much larger than C_{te}.

$$C_e = C_{De} + C_{Te}$$

$$C_e \approx C_{De}$$

It can be shown that

$$C_e = \frac{g_m}{2\pi f_T}$$

The short circuit current gain at high frequencies

The equivalent circuit of a common emitter amplifier using the hybrid-π circuit and having a load resistance R_L is shown in Fig 10.2. This circuit can be simplified to that shown in Fig 10.3 for the purpose of calculating the short-circuit current gain.

The load current I_L is given by

$$I_L = -g_m V_{b'e}$$

where

$$V_{b'e} = \frac{I_i}{g_{b'e} + j\omega(C_e + C_c)}$$

The short-circuit current gain is given by

$$A_i = \frac{I_L}{I_i} = \frac{-g_m}{g_{b'e} + j\omega(C_e + C_c)}$$

Using the results given earlier, we can write

$$A_i = \frac{-h_{fe}}{1 + j(f/f_\beta)}$$

where

$$|A_i| = \frac{h_{fe}}{\left[1 + (f/f_\beta)^2\right]^{1/2}}$$

and

$$f_\beta = \frac{g_{b'e}}{2\pi(C_e + C_c)}$$

$$= \frac{1}{h_{fe}} \frac{g_m}{2\pi(C_e + C_c)}$$

When $f = f_\beta$, $|A_i|$ is $1/\sqrt{2}$ of its low frequency value h_{fe}. The range of frequencies up to f_β is known as the bandwidth of the circuit.

The frequency f_T

An important parameter is the frequency f_T, which is the frequency at which the amplitude of the short-circuit common emitter current gain becomes unity.

Since

$$h_{fe} >> 1$$

$$f_T \approx h_{fe} f_\beta$$

and

$$A_i = -\frac{h_{fe}}{1 + jh_{fe}(f/f_T)}$$

A graph showing the variation of short-circuit current gain with frequency is shown in Fig 10.4.

Short-circuit current gain-bandwidth product

The frequency f_T is the product of the low frequency current gain h_{fe} and the upper 3 dB frequency. If two transistors having the same f_T are available, the one with the lower h_{fe} will have the larger bandwidth.

Vocabulary

available	vorhanden, erhältlich *adj*	**hybrid-π circuit**	Hybridschaltung *f*
behaviour	Verhalten, Benehmen *n*	**instantaneous**	sofort *adj*
data sheet	Datenblatt *n*	**magnitude**	Größe *f*
decrease	verkleinern, abnehmen *v*	**result from**	sich ergeben *v*
gradual	allmählich *adj*	**simplify**	vereinfachen *v*

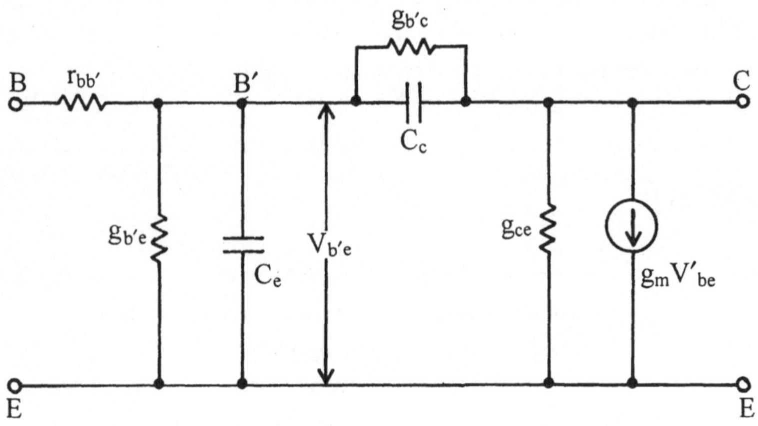

Fig 10.1 The common emitter hybrid-π circuit

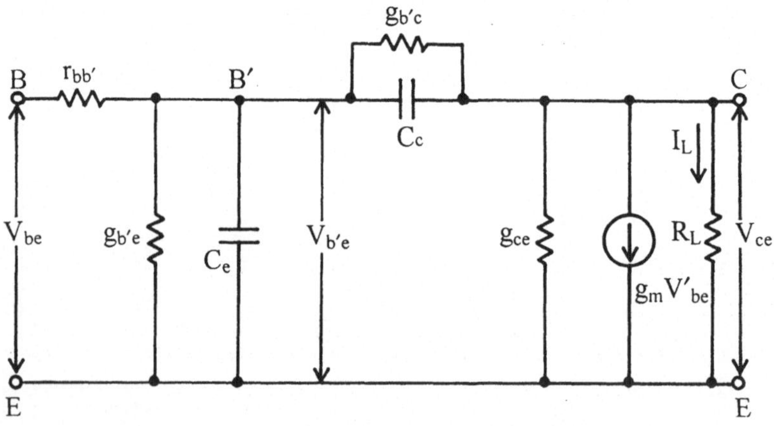

Fig 10.2 The hybrid-π circuit with a resistive load

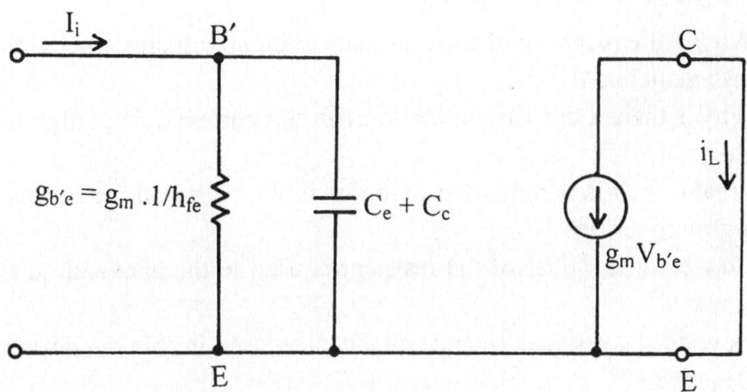

Fig 10.3 Simplified version of the hybrid-π circuit used for calculating the short-circuit current gain

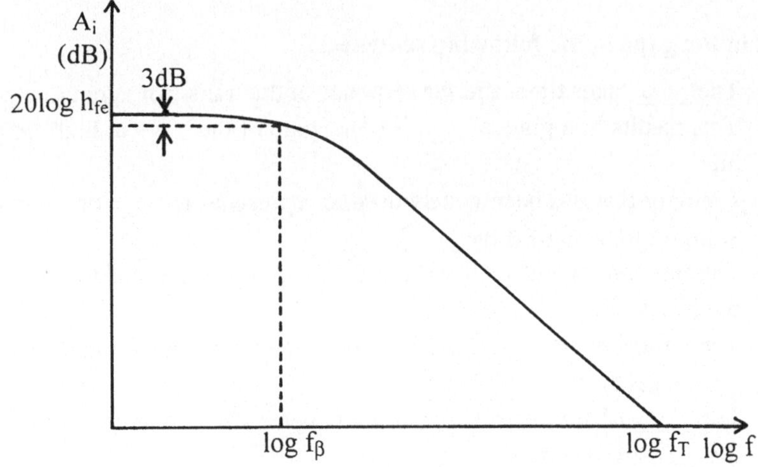

Fig 10.4 Variation of the CE short-circuit current gain with frequency

Exercises X

1. Answer the following questions:

a) Why is the response of the transistor to input voltages and currents not instantaneous ?
b) Why is there a decrease in the short circuit current gain at high frequencies ?
c) What circuit is widely used to represent the behaviour of a transistor at high frequencies ?
d) How can the values of the resistances used in the above circuit be obtained ?
e) How do the resistances and capacitances used in this circuit vary with frequency ?
f) What is the range of frequencies up to the upper 3 dB frequency called ?
g) What happens at the frequency f_T ?
h) How is f_T related to the low frequency current gain h_{fe} ?
i) How can the value of C_C be found ?
j) Which two capacitances add up to form the capacitance C_C ?

2. Fill in the gaps in the following sentences:

a) The ____ takes time, and the response of the transistor is not ____.
b) This results in a gradual ____ in the value of the ____ at high frequencies.
c) A circuit that has been widely used to represent the ____ of a transistor at high frequencies is the ____.
d) The resistances and ____ used in this circuit are assumed to be ____ of the frequency.
e) The values of the ____ in a hybrid-π circuit can be obtained from the ____ h-parameters.
f) The ____ of frequencies up to the upper 3 dB frequency is known as the ____ of the circuit.
g) At the frequency f_T the ___ of the short-circuit current gain becomes ____.
h) The frequency f_T is the product of the ____ and the ____ frequency.

i) The currents flowing through a transistor result from the ____ of charge carriers between the ____ of the transistor.
j) The capacitance C_C is the ____ of the collector to base ____capacitance with open input.

3. Translate into English:

a) Die durch einen Transistor fließenden Ströme beruhen auf der Diffusion von Ladungsträgern zwischen den verschiedenen Zonen des Transistors. Da der Diffusionsprozeß eine bestimmte Zeit benötigt, ist die Ansprache des Transistors auf Änderungen der Eingangsspannungen und -ströme nicht unmittelbar.
b) Der Betrag der Kurzschlußstromverstärkung der Emitterschaltung nimmt mit der Frequenz ab. Also muß das zur Darstellung des Hochfrequenzverhaltens des Transistors verwendete Ersatzschaltbild sich von dem bei niedrigen Frequenzen verwendeten Ersatzschaltbild unterscheiden.
c) Ein Ersatzschaltbild, das häufig zur Darstellung des Verhaltens des Transistors bei hohen Frequenzen verwendet wird, ist das Hybridschaltbild. Die in dieser Schaltung verwendeten Widerstände und Kapazitäten sind frequenzunabhängig.
d) Die Werte der Widerstände und Leitwerte können aus den Niederfrequenz h-Parametern ermittelt werden. Die Kapazität C_C ist der bei eingangsseitigem Leerlauf gemessene Wert der Kollektor-Basis Kapazität.
e) Ein wichtiger Parameter ist die Transitfrequenz f_T, welche diejenige Frequenz ist, bei der die Kurzschlußstromverstärkung der Emitterschaltung zu eins wird. Ein den Verlauf der Kurzschlußstromverstärkung über der Frequenz darstellenden Graphen zeigt Abb. 10.4.

11 Feedback amplifiers

The characteristics of an amplifier can be modified and improved considerably by the use of negative feedback. In this process a part of the output signal is combined with the input signal. If the output signal increases in magnitude, then the effect of the negative feedback signal is such that it opposes the increase in the output signal.

Advantages of negative feedback

- The gain of an amplifier varies for many reasons. Transistors of the same type may have parameters which vary by as much as 50%. Changes in temperature and other factors can cause large changes in the gain of an amplifier. The application of negative feedback tends to stabilize the amplifier and reduce variations in gain due to any cause.
- Distortion, noise, hum, etc, which are produced in the amplifier are reduced.
- The input and output impedances of an amplifier can be increased or decreased by the application of negative feedback.
- The frequency response and linearity of an amplifier can be improved. The gain of an amplifier usually decreases at high frequencies. The application of negative feedback increases the bandwidth at the expense of gain as shown in Fig 11.1.

Disadvantages of negative feedback

- The midfrequency gain of an amplifier is considerably reduced by the application of negative feedback. The improvement in the characteristics of the amplifier are obtained at the expense of the overall gain of the amplifier.
- The application of negative feedback may also result in the amplifier becoming unstable and breaking into oscillations. This is due to the fact that the phase shift of the signal produced by the amplifier varies with frequency, and a feedback signal that is negative at one frequency can become positive at another frequency.

Amplifier with voltage series-feedback

Many types of feedback circuits are possible, of which voltage-series feedback, current series-feedback, voltage-shunt feedback, and current-shunt feedback are some common types.

The block diagram of an amplifier with voltage-series feedback is shown in Fig 11.2. The voltage gain without feedback is A. A fraction β of the output signal is added to the input signal. Here the polarity of the amplifier gain and the feedback are considered to be positive. The consequences of making β positive or negative can be considered later.

The gain without feedback A is given by

$$A = \frac{V_0}{V_i}$$

The input signal V at the input of the amplifier after feedback is applied is

$$V = V_i + \beta V_0$$

The output voltage becomes

$$V_0 = AV$$

or

$$V_0 = A(V_i + \beta V_0)$$

$$V_0(1 - \beta A) = AV_i$$

The gain with feedback A_f is therefore given by

$$A_f = \frac{A}{1 - \beta A}$$

For negative feedback, β is negative and the expression becomes

$$A_f = \frac{A}{1 + \beta A}$$

The gain of an amplifier is usually high and $\beta A >> 1$. In this case A_f becomes

$$A_f = \frac{1}{\beta}$$

It will be seen that the gain of the amplifier with feedback, depends only on the feedback factor β and is independent of variations in A.

The gain of a voltage amplifier with voltage-series feedback is reduced by a factor of $(1+\beta A)^{-1}$. It can be shown that the distortion as well as the variation in gain are also reduced by the same factor. A numerical example is given below.

Example: A voltage amplifier has a gain of A = 10000 and a maximum variation in gain of 50%. Find the gain, and the maximum variation in gain when negative series-voltage feedback is applied, given that the feedback factor β = 1/100.

The gain becomes

$$A_f = \frac{A}{1+\beta A} = \frac{10000}{1+100}$$

$$A_f \approx 100$$

The maximum variation in gain is also reduced by the same factor of 100

Maximum variation in gain $\approx \frac{50}{100}\% = 0.5\%$, or $A_f = 100 \pm 0.5\%$

Vocabulary

application	Anwendung *f*	**modify**	modifizieren, umwandeln *v*
cause	verursachen *v*	**noise**	Geräusch *n*
consequence	Folge, Wirkung *f*	**oscillation**	Schwingung *f*
considerably	wesentlich, beträchtlich *adj*	**phase**	Phase *f*
expense	Aufwand *m*	**produce**	erzeugen *v*
feedback	Rückkopplung *f*	**reduce**	verkleinern *v*
fraction	Bruchteil *m*	**shunt**	parallelschalten *v*
hum	Brummen *n*	**stabilize**	Konstant halten, stabilisieren *v*
improve	verbessern *v*	**unstable**	instabil *adj*

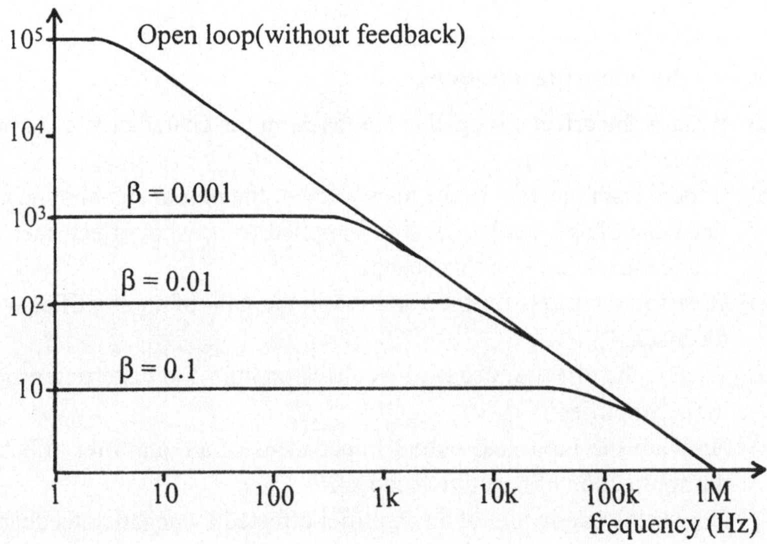

Fig 11.1 Frequency response of an amplifier for different feedback factors

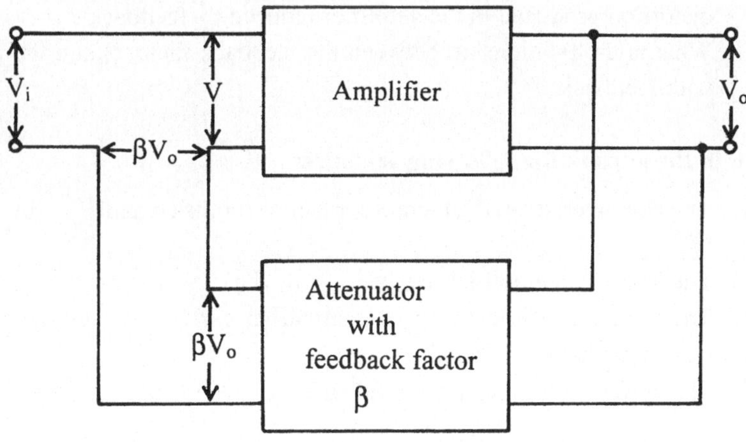

Fig 11.2 Block diagram of an amplifier with voltage series-feedback

Exercises XI

1. Answer the following questions

a) What is the effect of negative feedback on the characteristics of an amplifier ?
b) If there is an increase in the magnitude of the output signal of an amplifier which has negative feedback applied to it, what effect does negative feedback have on this change ?
c) Give some reasons for the variation in the gain of an amplifier without feedback.
d) What is the effect of negative feedback on hum and distortion produced in the amplifier ?
e) How are the input and output impedances of an amplifier affected by the application of negative feedback ?
f) How is the bandwidth of an amplifier affected by negative feedback ?
g) What is the effect of negative feedback on the midfrequency gain of an amplifier ?
h) What is the effect of negative feedback on the stability of an amplifier ?
i) An amplifier has a negative feedback factor of β. By what factor is the distortion produced in the amplifier reduced by feedback ?
j) What is the relationship between the feedback factor β and the gain A_f with feedback ?

2. Fill in the gaps in the following sentences

a) The characteristics of an amplifier can be modified and ____ by the use of ____.
b) The gain of an amplifier can be ____ by the ____ of negative feedback.
c) The input and output ____ of an amplifier can be ____ by the application of negative feedback.
d) The application of negative feedback ____ bandwidth at the expense of ____.
e) The ____ gain of an amplifier is ____ by the application of negative feedback.
f) The application of negative feedback may result in the amplifier becoming ____ and breaking into ____.

g) The ____ of an amplifier usually ____ at high frequencies.
h) The phase ____of the amplified signal ____ with frequency.
i) Variations in ____, and changes in the values of the ____ can cause large changes in the gain of an amplifier.
j) Distortion which is ____ in an amplifier can be ____ by negative feedback.

3. Translate into English:

a) Die charakteristischen Eigenschaften eines Verstärkers können durch Gegenkopplung verbessert werden. Bei diesem Prozeß wird ein Teil des Ausgangssignals dem Eingangssignal zugeführt.
b) Veränderungen der Temperatur und anderer Faktoren wie Transistorparameter können starke Veränderungen der Verstärkung eines Verstärkers verursachen. Transistoren gleichen Typs können Abweichungen bis zu 50% in ihren Parametern haben.
c) Der Einsatz von Gegenkopplung stabilisiert den Verstärker und verringert Variationen der Verstärkung jedweder Ursache. Verzerrungen, Rauschen und Brummen werden ebenfalls verringert.
d) Die Ausgangs- und Eingangsimpedanz eines Verstärkers können durch Einsatz von Gegenkopplung erhöht oder verringert werden. Die Bandbreite des Verstärkers wird auf Kosten der Verstärkung erhöht.
e) Der Einsatz von Gegenkopplung kann die Instabilität des Verstärkers verursachen. Dieses beruht darauf, daß die Phasenverschiebung des Ausgangssignals sich über der Frequenz ändert.

12 Operational amplifiers

The operational amplifier in integrated circuit form has been widely used as a building block in analog circuits. It has the advantages of versatility, reliability, small size, low cost, and the ability to amplify D.C signals.

The operational amplifier is a multistage direct coupled high gain amplifier, whose overall response can be controlled by the addition of feedback. For most practical purposes one can assume that an operational amplifier has

- infinite voltage gain (typical value $2x10^5$)
- infinite input impedance (typical value 2 MΩ)
- zero output impedance (typical value 75 Ω)
- large bandwidth (typical value a few MHz)

The basic operational amplifier

The block schematic diagram of an operational amplifier is shown in Fig 12.1. It has two input terminals and one output terminal. The voltages V_1 and V_2 are applied to the two input terminals which are called the inverting and noninverting terminals respectively. The noninverting voltage gain V_0/V_2 is positive, while the inverting voltage gain V_0/V_1 is negative. When $V_1 = V_2$, the output voltage should theoretically be zero. A single-ended amplifier can be constructed by earthing one of the input terminals.

The inverting amplifier

One of the most commonly used operational amplifier circuits is the basic inverting circuit shown in Fig 12.2. In this circuit feedback is obtained by using the two impedances Z and Z'. If the input impedance R_i of the amplifier is assumed to be infinity, no current flows into the amplifier input terminals. Therefore the same current I that flows through Z, also flows through Z'.

Since A_V is very large, the potential difference between the input terminals is small. From this it follows that both input terminals are virtually at the same potential, and that the inverting terminal is at earth potential. It can be assumed that a virtual short circuit or a virtual earth exists at the input. The term virtual implies that no current flows into this short circuit.

$$A_f = \frac{V_0}{V_S} = -\frac{IZ'}{IZ}$$

$$A_f = -\frac{Z'}{Z}$$

The noninverting amplifier

In a noninverting circuit the output voltage is in phase with the input voltage. The basic circuit has voltage-series feedback as shown in Fig 12.3. The feedback voltage is V_2. If $I_2 = 0$, the feedback factor β becomes

$$\beta = \frac{V_2}{V_0} = \frac{Z}{Z + Z'}$$

As has been previously shown

$$A_f = \frac{1}{\beta} \qquad \text{when } \beta A >> 1$$

$$A_f = \frac{Z + Z'}{Z} = 1 + \frac{Z'}{Z}$$

The application of voltage-series feedback, results in an amplifier with a very high input impedance and a very low output impedance.

Drift in D.C amplifiers

One of the disadvantages of a D.C amplifier is that any changes in D.C operating conditions can cause voltage changes that are indistinguishable from the signal being amplified. Such voltage changes can occur for example as a result of temperature variations and are termed drift voltages.

Differential amplifiers

A good way of reducing amplifier drift is to use a balanced circuit, where the voltage changes in one part of the circuit are balanced by equal and opposite changes in another part of the circuit. An example of such a balanced circuit is the differential amplifier shown in Fig 12.4. As the name implies, the output voltage is proportional to the voltage difference between the two input voltages. If two voltages V_1 and V_2 are applied to the two inputs, the output voltage V_0 in an ideal amplifier is given by

$$V_0 = A_{VD}(V_1 - V_2)$$

where A_{VD} is the differential voltage gain of the amplifier. However due to imbalance in the amplifier, there is usually an output signal V_0 even when the same voltage is applied to both input terminals. This type of input is called the common mode input, and the corresponding gain is called the common mode gain A_{VCM}. It is desirable that the common mode gain be as small as possible.

The common mode rejection ratio (CMRR)

A criterion of the quality of a differential amplifier is its common mode rejection ratio (CMRR). This is expressed in dB as,

$$\text{CMRR} = 20\log_{10}\left(\frac{A_{VD}}{A_{VCM}}\right)\text{dB}$$

For the simple amplifier shown in Fig 12.4, it can be shown that

$$A_{VCM} = -\frac{R_1}{2R_3}$$

To improve the CMRR, A_{VCM} should be reduced, and this can be done by increasing R_3. A better way of doing this is to replace R_3 by a transistor as a constant current source. A modified circuit using a transistor instead of R_3 is shown in Fig 12.5.

Offset error voltages and currents

An ideal operational amplifier shows perfect balance. This means that the output voltage is zero, when the voltages at both inputs are zero. In practice however, the two transistors in the input stage of an operational amplifier are not perfectly matched, and this causes some imbalance. The amplifier can be balanced by applying an offset voltage between the two input terminals. D.C error voltages and currents can be measured at the input and output terminals, and these form part of the specifications normally provided by the manufacturer.

Vocabulary

ability	Fähigkeit *f*	**match**	anpassen *v*
advantage	Vorteil *m*	**modify**	ändern, modifizieren *v*
assume	annehmen *v*	**occur**	vorkommen, eintreten *v*
balance	ausgleichen *v*	**offset voltage**	Gegenspannung *f*,
block schematic diagram	Blockschaltbild *n*	**operational amplifier**	Operationsverstärker *m*
CMRR	Gleichtaktunterdrückung *f*	**overall**	gesamt *adj*
control	beherrschen, kontrollieren *v*	**performance**	Leistungsfähigkeit *f*
D.C (direct current)	Gleichstrom *m*	**possible**	möglich *adj*
describe	beschreiben *v*	**practical**	praktisch *adj*
differential input	Differenzeingang *f*	**purpose**	Zweck *f*,
drift voltage	Driftspannung *f*	**reliability**	Zuverlässigkeit *f*
error voltage	Fehlerspannung *f*	**specification**	Richtlinie, Vorschrift *f*
imbalance	Ungleichgewicht *n*	**typical value**	üblicher Wert *m*
indistinguishable	unmerklich *adj*	**versatility**	Vielseitigkeit *f*
in practice	in der Praxis *f*	**virtually**	praktisch, eigentlich *adv*
inverting amplifier	invertierender Verstärker *m*	**widely**	weit *adv*
inverting terminal	invertierender Eingang *m*		

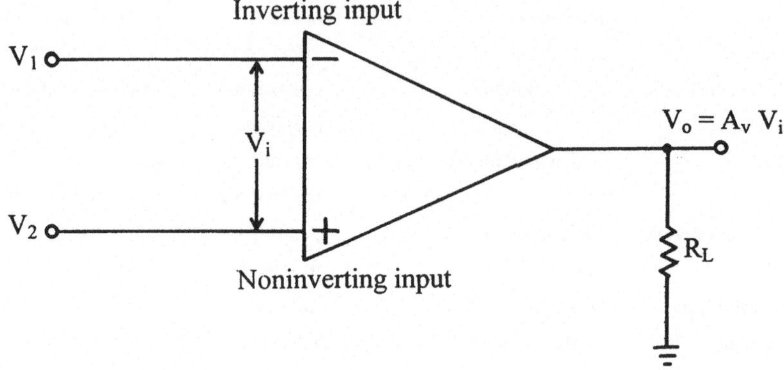

Fig 12.1 Basic operational amplifier

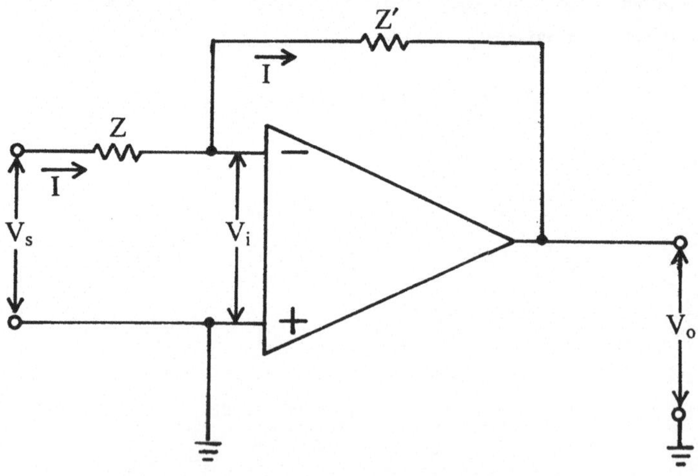

Fig 12.2 Inverting operational amplifier

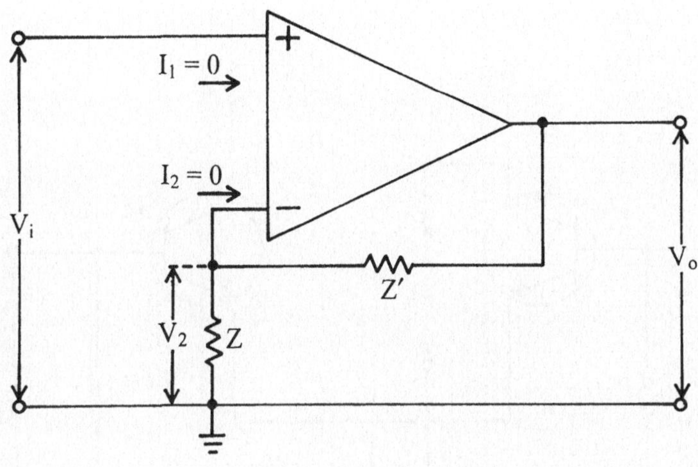

Fig 12.3 Noninverting operational amplifier circuit

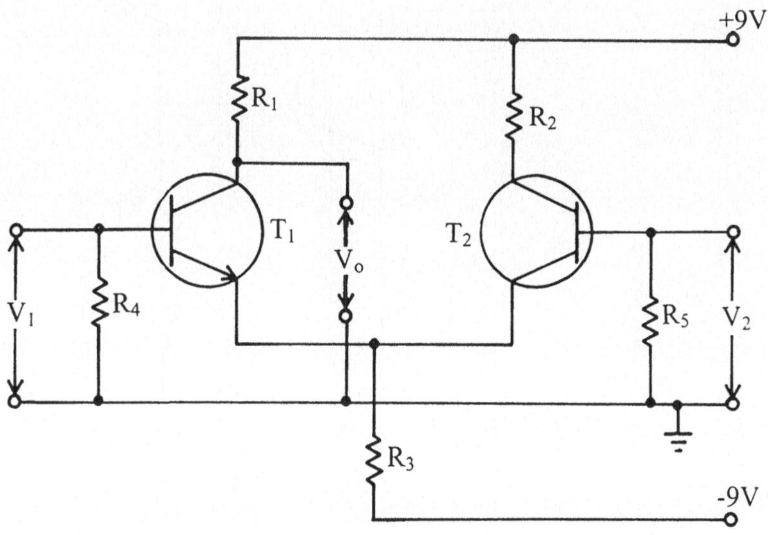

Fig 12.4 Basic differential amplifier circuit

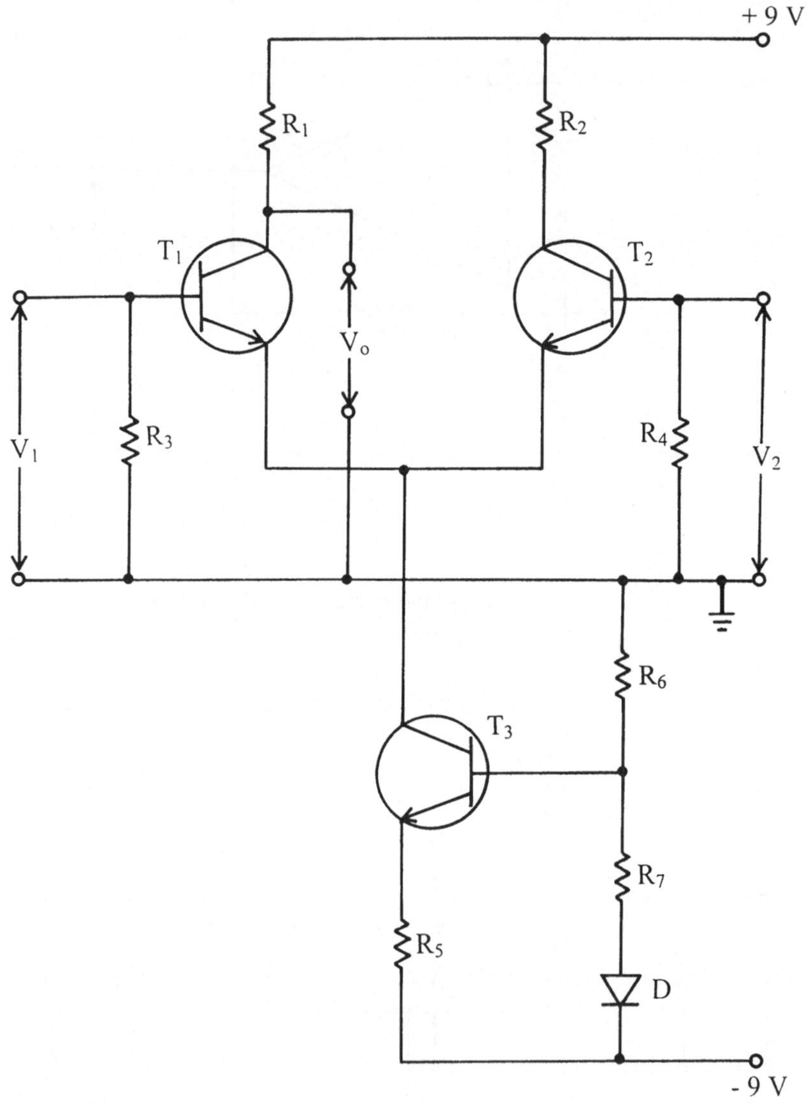

Fig 12.5 Differential amplifier with transistor as constant current source

Exercises XII

1. Answer the following questions:

a) What advantages do operational amplifiers in integrated circuit form have ?
b) What values of voltage gain and bandwidth does a typical operational amplifier have ?
c) What are the two input terminals of an operating amplifier called ?
d) How can a single-ended operational amplifier be constructed ?
e) How is feedback applied to an inverting amplifier ?
f) What quantities affect A_f, the gain with feedback of an operational amplifier ?
g) What is the phase relationship between output and input voltages for a noninverting amplifier ?
h) How can drift in a D.C amplifier be reduced ?
i) What is the output voltage of a differential amplifier proportional to ?
j) What causes imbalance in a practical operational amplifier ?

2. Fill in the gaps in the following sentences:

a) The ____ operational amplifier has been widely used as a building block in ____.
b) An operational amplifier has ____ input terminals and one ____ terminal.
c) A ____ amplifier can be constructed by earthing one of the ____ .
d) If the input impedance of the amplifier is assumed to be ____ , no current flows into the amplifier ____.
e) Both input terminals are ____ at the same potential and the inverting terminal is at ____ potential.
f) In a ____ circuit the output voltage is in phase with the ____.
g) In a ____ circuit voltage changes in one part of the circuit are balanced by equal and ____ changes in another part of the circuit.
h) In a ____ amplifier, the output voltage is ____ to the difference between the input voltages.
i) Voltage changes can occur as a result of ____ variations and are termed ____ voltages.

j) The ____ of a differential amplifier may be assessed by measuring its ____.

3. Translate into English:

a) Der Operationsverstärker als integrierte Schaltung wird vielfach als Baustein in analogen Schaltungen eingesetzt. Er hat die Vorteile der Vielseitigkeit, des niedrigen Preises, der hohen Zuverlässigkeit und der Fähigkeit, Gleichstromsignale zu verstärken.
b) Der Operationsverstärker ist ein mehrstufiger, direkt gekoppelter Hochleistungsverstärker, dessen Gesamtverstärkung durch Rückkopplung reguliert werden kann.
c) In der Praxis kann man davon ausgehen, daß ein Operationsverstärker eine unendliche Spannungsverstärkung, eine unendliche Eingangsimpedanz, eine Ausgangsimpedanz von Null und eine große Bandbreite besitzt.
d) Eine gute Möglichkeit, die Drift des Verstärkers zu verkleinern, ist der Einsatz einer symmetrischen Schaltung, in der die Spannungsänderungen in einem Teil der Schaltung durch gleichgroße entgegengesetzte Änderungen in einem anderen Teil der Schaltung ausgeglichen werden.
e) Ein idealer Operationsverstärker weist perfekte Symmetrie auf, d.h., daß die Ausgangsspannung Null ist, wenn die Spannungen an beiden Eingängen Null sind. Normalerweise sind die Transistoren der Eingangsstufe nicht perfekt gepaart, und dies verursacht eine gewisse Unsymmetrie.

13 Linear analog systems

The integrated circuit (IC) operational amplifier has been used as a building block in the construction of a variety of linear and nonlinear analog systems.

Linear analog systems

Circuits containing operational amplifiers and a few discrete components can perform many mathematical operations. This is the reason for the use of the name operational amplifier. A few circuits of this type are discussed below.

Multipliers and dividers

It has been seen that the voltage gain for an inverting amplifier with feedback is given by

$$A_f = -\frac{Z'}{Z} = k$$

The output voltage is k times the input voltage. Almost any value of k can be obtained by choosing suitable values of resistors for Z and Z'. Such a circuit can be used as a multiplier or a divider.

Phase shifter

A phase shifter can be constructed using the inverting amplifier as a basis. The values of Z and Z' are such that they are equal in magnitude but different in phase angle. Any phase shift from 0° to 360° (± 180°) may be obtained.

Adder or subtractor

The circuit shown in Fig 13.1 can be used to add a number of input voltages v_1 to v_n.

Here $$I = \frac{v_1}{R_1} + \frac{v_2}{R_2} + \cdots + \frac{v_n}{R_n}$$

The output voltage $$v_0 = -R'I = -\left(\frac{R'}{R_1}v_1 + \frac{R'}{R_2}v_2 + \cdots + \frac{R'}{R_n}v_n\right)$$

If $$R_1 = R_2 = \cdots = R_n$$

then $$v_0 = -\frac{R'}{R_1}\left(v_1 + v_2 + \cdots + v_n\right)$$

Integrator

The inverting amplifier can perform the mathematical operation of integration if a capacitance C is used for Z', and a resistance R for Z, as shown in Fig 13.2. The input voltage need not be sinusoidal.

Here $$i = \frac{v_i}{R}$$

and $$v_0 = \frac{q}{C} = -\frac{1}{C}\int i\,dt$$

$$v_0 = -\frac{1}{CR}\int v_i\,dt$$

The output voltage is equal to the integral of the input voltage.

Differentiator

In this case a capacitor is used for Z, and a resistor R for Z', as shown in Fig 13.3. It will be seen that

$$i = \frac{dq}{dt} = C\frac{dv_i}{dt}$$

$$v_0 = -iR = -CR\frac{dv_i}{dt}$$

The output voltage is proportional to the time derivative of the input voltage.

The analog computer

The circuits that have been described above like the differentiator, integrator, adder, etc, can be combined together to form an analog computer. By choosing suitable circuits and component values an analog computer can be used to solve differential equations.

Active filters

Filters are often used for the purpose of restricting the frequency bandwidth of a circuit. Passive filters use a combination of inductances, capacitances, and resistances to achieve the necessary frequency response. Inductances have disadvantages in that they are expensive, and tend to pick-up hum from the mains.

The use of inductances can be avoided by using active filters. It is possible to simulate the behaviour of an LCR filter by using a circuit that contains only resistors and capacitors together with an operational amplifier. This eliminates the need for inductances.

The circuit of an RC bandpass filter is shown in Fig 13.4, and its frequency response in Fig 13.5. A flat-topped bandpass characteristic can be obtained by using two such filters in cascade (i.e. the output of the first filter feeding the input of the second). The resonant frequencies are slightly staggered (i.e. made slightly different). The response curve is shown in Fig 13.6.

Other linear analog systems

Many other linear analog systems using operational amplifiers as building blocks are available. Among these are voltage to current converters, amplifiers of various types, delay equalizers, etc.

Vocabulary

active filter	aktiver Filter *m*	**flat-topped characteristic**	Rechteckkurve *f*
analog system	Analogsystem *n*	**multiplier**	Multiplizierer *m*
avoid	vermeiden *v*	**nonlinear**	nichtlinear *adj*
bandpass filter	Bandfilter *n*, Bandpaß *m*	**obtain**	erhalten *v*
basic	grundlegend *adj*	**operation**	Vorgang *m*, Operation *f*
cascade	Kaskadenschaltung *f*	**perform**	leisten *v*
connection	Verbindung *f*	**phase shift**	Phasenverschiebung *f*
converter	Umwandler *m*	**pick up**	aufnehmen *v*
delay equalizer	Laufzeitentzerrer *m*	**reason**	Grund m, Ursache *f*
derivative	Differentialquotient *m*	**resistor, resistance**	Widerstand *m*
disadvantage	Nachteil *m*	**restrictive**	einschränkend *adj*
discuss	diskutieren, besprechen *v*	**simulate**	nachbilden, simulieren *v*
discrete	einzeln, diskret *adj*	**sinusoidal**	sinusformig *adj*
divider	Teiler *m*, Dividierer *m*	**staggered**	versetzt *adj*

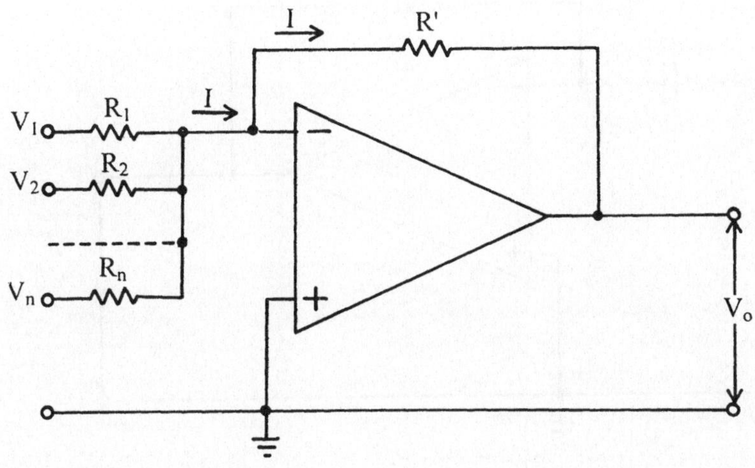

Fig 13.1 An operational adder

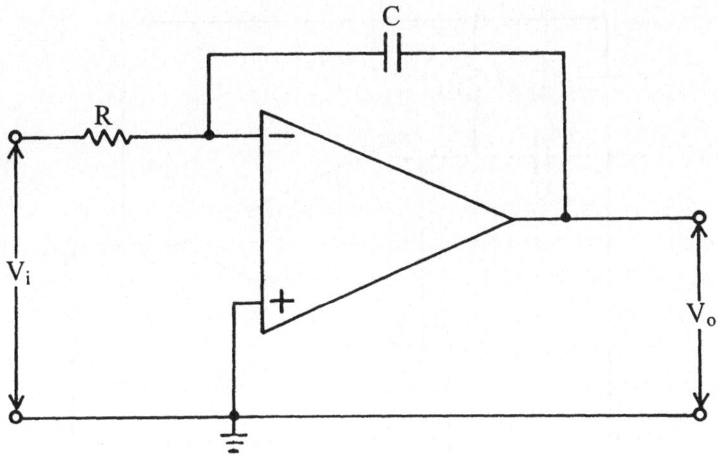

Fig 13.2 An operational integrator

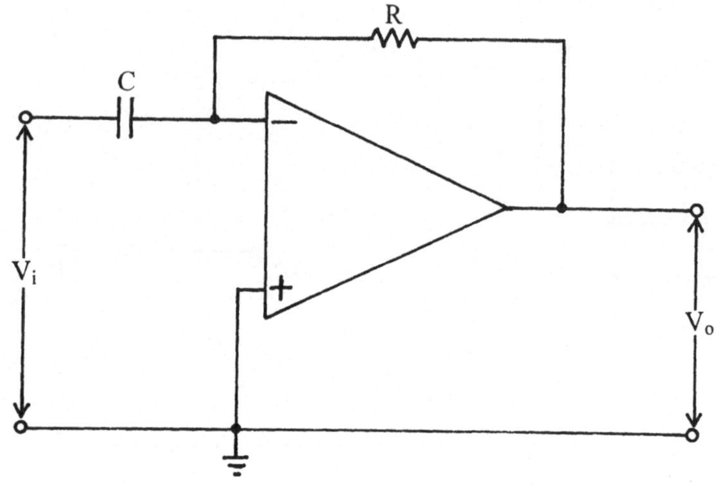

Fig 13.3 An operational differentiator

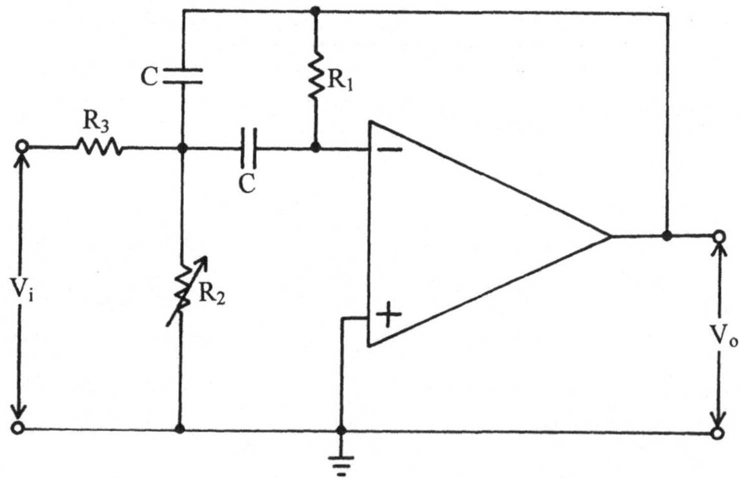

Fig 13.4 A tuned active RC filter

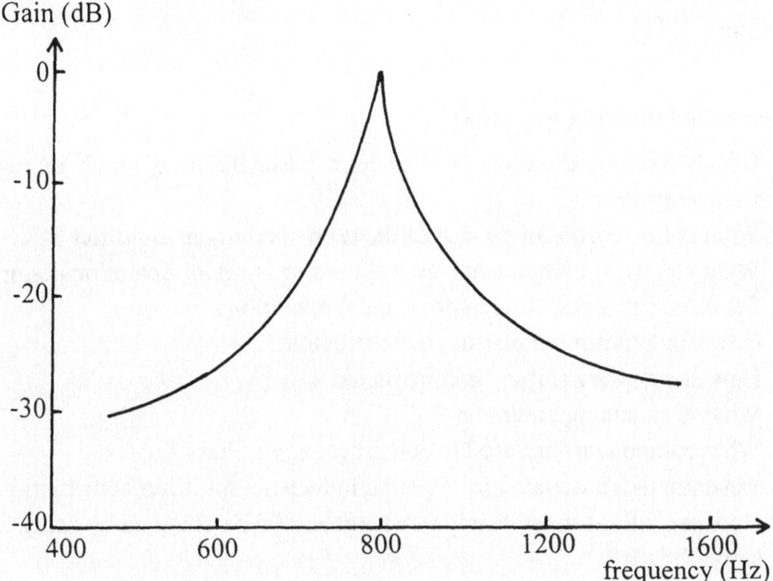

Fig 13.5 Frequency response of the tuned active filter

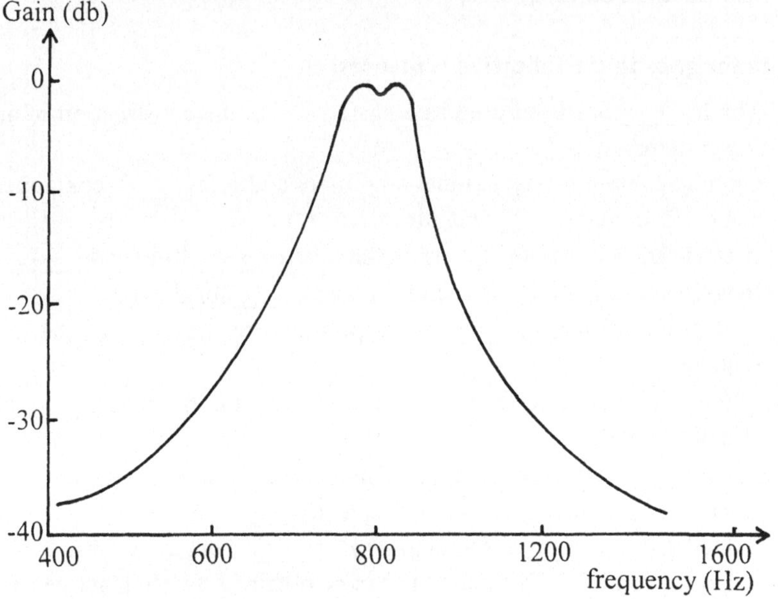

Fig 13.6 Frequency response of two cascaded filters with staggered resonant frequencies

Exercises XIII

1. Answer the following questions

a) Which integrated circuit is used as a basic building block in many analog systems ?
b) What is the reason for the use of the term operational amplifier ?
c) What electrical components are required to build an operational amplifier circuit that performs mathematical operations ?
d) How can a multiplier circuit be constructed ?
e) How can a phase shifter be constructed ?
f) What is an analog computer ?
g) What components are used to construct active filters ?
h) What are the disadvantages of using inductances in filter construction?
i) How can a flat-topped bandpass characteristic be obtained by using two active filters ?
j) Give the names of some linear analog systems which use operational amplifiers as building blocks.

2. Fill in the gaps in the following sentences:

a) The IC ____ has been used as a basic ____ in the construction of analog systems.
b) Circuits containing operational amplifiers and a few ____ components can perform many ____ operations.
c) A phase shifter can be ____ using the inverting amplifier as a ____.
d) In an integrator, a ____ is used for Z'and a ____ for Z.
e) In a differentiator, the ____ is proportional to the ____ of the input voltage.
f) Filters are often used for the purpose of ____ the frequency ____ of a circuit.
g) Passive ____ use a combination of inductances, capacitances, and resistances, to achieve the necessary frequency ____.
h) Inductances are ____ and tend to pick-up ____ from the mains.
i) A flat-topped ____ characteristic can be obtained by using active filters in ____.

j) By choosing suitable circuits and ____, an analog computer can be used to solve ____.

3. Translate into English:

a) Schaltungen, die Operationsverstärker und wenige diskrete Komponenten enthalten, können viele mathematische Operationen durchführen, wie Addition, Multiplikation, Differentiation und Integration.
b) In einer Multiplizierschaltung ist die Ausgangsspannung das k-fache der Eingangsspannung. Beinahe jeder Wert von k kann durch geeignete Wahl von Widerstandswerten für Z und Z' erreicht werden. Eine solche Schaltung kann auch als Dividierer eingesetzt werden.
c) Ein invertierender Verstärker kann die mathematische Operation der Integration ausführen, wenn eine Kapazität C für Z' und ein Widerstand R für Z eingesetzt werden. Dann ist die Ausgangsspannung gleich dem Integral der Eingangsspannung.
d) Schaltungen wie Differentiator, Integrator, Addierer etc. können zu einem analogen Rechner kombiniert werden. Ein solcher Rechner kann zum Lösen von Differentialgleichungen eingesetzt werden.
e) Die Verwendung von Induktivitäten kann durch den Einsatz von aktiven Filtern vermieden werden. Es ist möglich, das Verhalten eines LCR-Filters durch eine Schaltung nachzubilden, die lediglich Widerstände, Kapazitäten und einen Operationsverstärker enthält.

14 The field effect transistor

The operation of the junction transistor depends on the flow of both majority and minority charge carriers. The other widely used type of transistor is the field effect transistor whose operation depends only on the flow of majority carriers. The field effect transistor is therefore called a unipolar device.

There are two types of field effect transistors:

- The junction field effect transistor (JFET or FET)
- The metal oxide field effect transistor (MOSFET or MOS), also called the insulated gate field effect transistor (IGFET)

The field effect transistor has many advantages over the junction transistor. It is simpler to fabricate and occupies less space in an integrated circuit. It is less noisy and has a high input impedance. Its main disadvantage is that it has a smaller gain-bandwidth product than the junction transistor and is therefore rarely used in amplifier circuits. Its main application is in large scale integrated (LSI) digital circuits such as semiconductor memories.

The junction field effect transistor

Junction field effect transistors are available as both n-channel and p-channel types. The construction of an n-channel FET is shown in diagrammatic form in Fig 14.1. A diagram showing the circuit symbol and the necessary D.C voltages is given in Fig 14.2. The n-channel FET is constructed from a bar of n-type silicon which has an ohmic metal contact at each end. The contacts at the ends of the bar are called the source and drain. On both sides of the bar, two heavily doped regions of p^+ type silicon are formed creating p-n junctions. These p^+ type regions are joined together and called the gate.

If a D.C voltage is applied between source and drain, there is a steady flow of electrons from source to drain. The flow of electrons can be controlled by applying a potential difference between gate and source.

When a voltage is applied between gate and and source in such a direction as to reverse bias the p-n junctions, depletion regions are formed as shown in Fig

14.1, and the flow of current is restricted to the channel between these regions. As the magnitude of the gate to source voltage V_{GS} is increased, the channel becomes narrower and the current is reduced.

The ohmic voltage drop along the bar causes the width of the conducting channel to decrease as we move along the bar. The channel cannot close completely but remains at some small but finite width.

The output characteristics and the pinch-off voltage

The output or drain characteristics of an n-channel FET are shown in Fig 14.3. To explain the shape of these characteristics consider one of the curves for which V_{GS} is constant. If V_{DS} is increased from zero, I_D increases almost linearly with V_{DS}. With increasing current the conducting channel decreases in width. Eventually a value of V_{DS} called the pinch-off voltage is reached.

Beyond this is the saturation region, where increasing V_{DS} does not cause an increase in I_D. Beyond the saturation region there is an avalanche breakdown region corresponding to higher values of V_{DS}.

The metal oxide field effect transistor (MOS or MOSFET)

The metal oxide field effect transistor is different in construction from the junction field effect transistor. Two types of MOSFET are available, the enhancement type and the depletion type.

The enhancement MOSFET

The construction of a p-channel enhancement MOSFET is shown in Fig 14.4. It has a lightly doped n-type substrate which has two highly doped p^+ regions formed in it by diffusion. The p^+ regions act as the source and drain. The gate is a metal electrode insulated from the silicon bar by a layer of silicon dioxide. There are two junctions p-n and n-p between the source and the drain. If a potential difference is applied between the source and the drain, no current flows whatever the polarity of the applied voltage. This is because one of the junctions is always reverse biased.

If the gate is now made negative with respect to the source, p-type carriers are attracted into the region between source and drain. A current can now flow from the p-type source to the p-type drain, and it can be said that the gate voltage causes an enhancement current to flow. Hence the name enhancement

MOSFET. The drain or output characteristics for a p-channel enhancement MOSFET are shown in Fig 14.5.

The depletion MOSFET

In the depletion MOSFET, an impurity of the same type as present in the source and drain is permanently diffused into the region between the source and the drain at the time of manufacture. In this way a permanent channel is formed, and a current can flow from source to drain even when no voltage is applied between gate and source. The construction of an n-channel depletion MOSFET is shown in Fig 14.6.

The current flow is due to the presence of majority carriers which in this case are electrons. If a negative gate voltage is applied, positive charges are induced in the channel and this causes a depletion in the number of majority carriers and in the drain current. Hence the use of the term depletion MOSFET. The depletion region near the drain is larger than that nearer the source. This is similar to the pinch-off effect that occurs in the JFET.

The depletion MOSFET can also be operated in the enhancement mode. For this purpose a positive voltage has to be applied to the gate so that negative charges are induced in the n-type channel. The drain or output characteristics for a depletion MOSFET showing both the depletion and enhancement modes is shown in Fig 14.7.

Vocabulary

bar	Stab *m*	**gain-bandwidth product**	Verstärkungsbandbreite-produkt *n*
bipolar	bipolar *adj*	**gate**	Gate (-Anschluß) *n*
compact	kompakt, gedrängt *adj*	**include**	enthalten, einschließen *v*
completely	vollständig *adv*	**induced channel**	induzierter Kanal *m*
contact	Anschluß, Kontakt *m*	**insulated gate FET**	FET mit isoliertem Gate *m*
control electrode	Steuerelektrode *f*	**junction FET**	Sperrschicht-FET
depend	abhängen v	**layer**	Schicht *f*
depletion MOSFET	Verarmungs-MOSFET *m*	**narrow**	eng *adj*
depletion mode operation	Verarmungsbetrieb *m*	**noisy**	rauschend *adj*
drain	Drain *m*	**occupy**	besetzen *v*
ease	Bequemlichkeit *f*	**pinch-off**	Abschnürung *f*
enhancement mode operation	Anreichungsbetrieb *m*	**restrict**	beschränken *v*
enhancement MOSFET	Anreichungs-MOSFET *m*	**source**	Source *f*
field effect transistor	Feldeffekttransistor *m*	**space**	Raum *m*
		unipolar	einpolig, unipolar *adj*

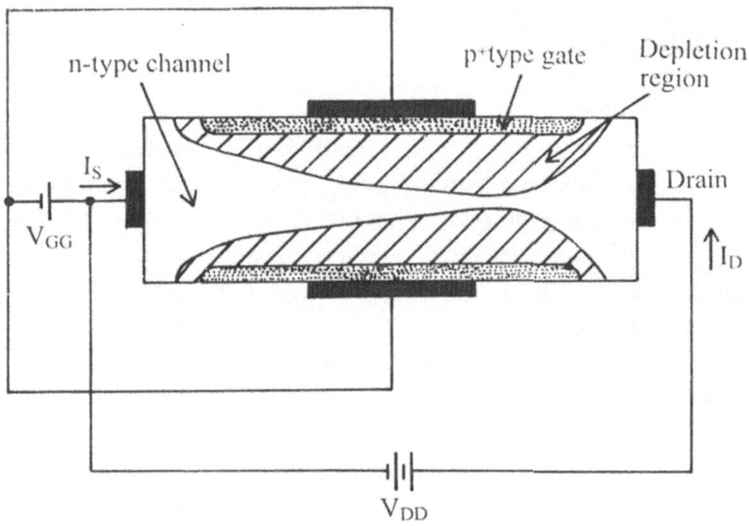

Fig 14.1 Diagram showing the n-type channel and the depletion region for an n-channel FET

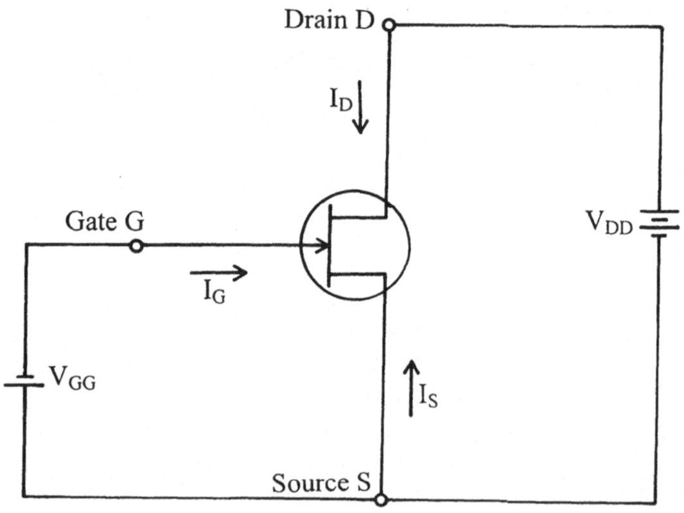

Fig 14.2 Basic circuit for an n-channel FET showing symbol and D.C voltages

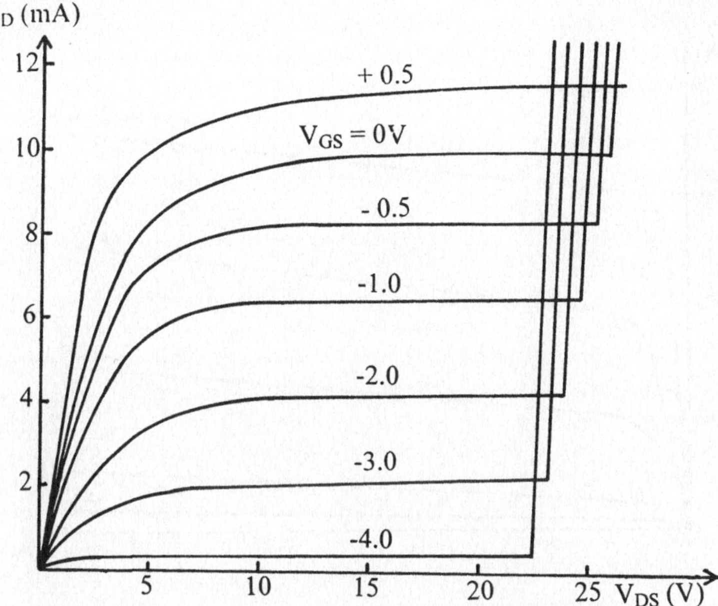

Fig 14.3 Output characteristics for an n-channel FET

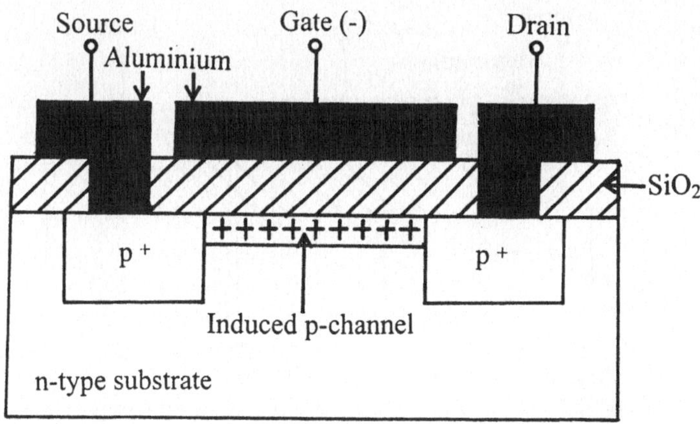

Fig 14.4 Enhancement or induced channel in a p-channel MOSFET

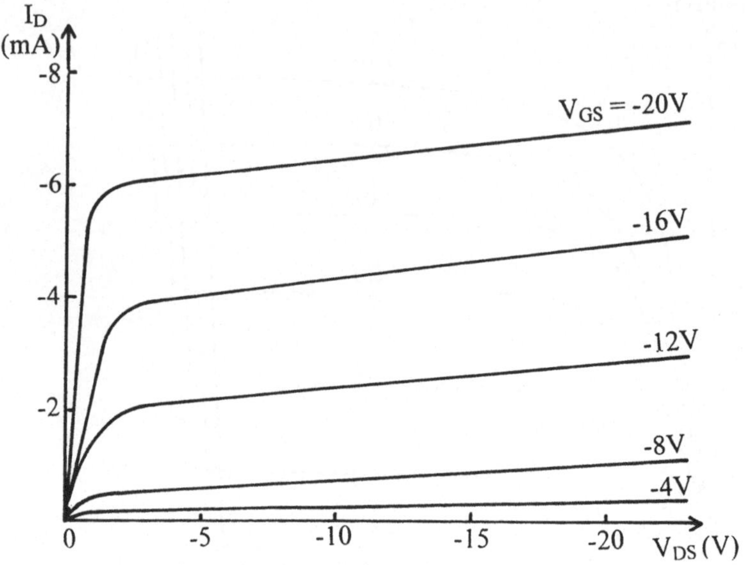

Fig 14.5 Drain or output characteristics for a p-channel enhancement MOSFET

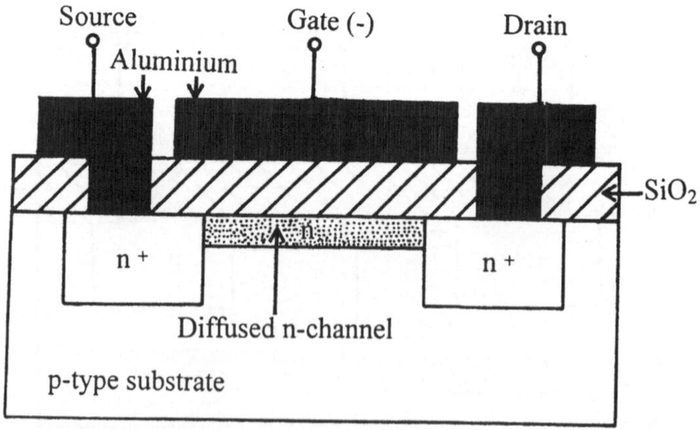

Fig 14.6 The construction of an n-channel depletion MOSFET

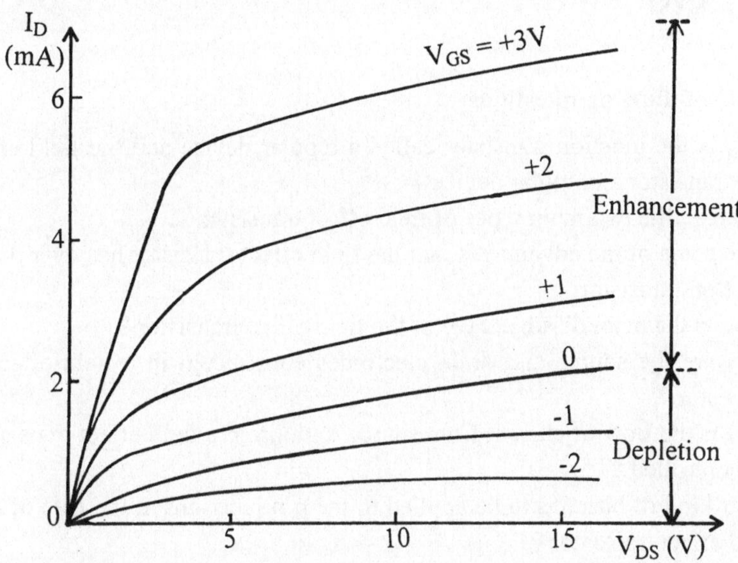

Fig 14.7 Drain or output characteristics for an n-channel depletion MOSFET

Exercises XIV

1. Answer the following questions:

a) Why is the junction transistor called a bipolar device and the field effect transistor a unipolar device ?
b) What are the two main types of field effect transistor ?
c) State some of the advantages that the field effect transistor has over the junction transistor.
d) What is the main disadvantage of the field effect transistor ?
e) How are the source and drain electrodes constructed in a field effect transistor ?
f) How is the flow of current from source to drain in a field effect transistor controlled ?
g) What kind of bias has to be applied to the p-n junctions at the gate of a field effect transistor ?
h) What is the difference between between the enhancement and depletion types of MOSFET ?
i) Which MOSFET can be operated in both the enhancement and depletion modes ?
j) Why is there no drain current in an enhancement MOSFET when $V_{GS} = 0$?

2. Fill in the gaps in the following sentences:

a) The ____ of the junction transistor depends on the flow of ____ charge carriers.
b) The field effect transistor is easier to ____ and occupies less space in an ____.
c) A field effect transistor is constructed from a ____ which has an ____ at each end.
d) The main ____ of the field effect transistor is that it has a smaller ____ than the junction transistor.
e) Majority carriers enter the bar through ____ and flow out through ____.
f) In a field effect transistor the flow of majority carriers can be ____ by applying a ____ between the gate and the source.

g) In a depletion MOSFET an impurity of the ____ as present in the source and drain is ____ into the region between the source and drain.
h) In ____ MOSFET no drain current flows unless a suitable potential difference is applied between ____.
i) In ____ MOSFET, the flow of current is due to ____ channel produced by applying a potential difference between ____.
j) A ____ MOSFET can also be operated in the ____ mode.

3. Translate into English:

a) Die Funktion von Bipolartransistoren hängt vom Fluß sowohl der Majoritäts- als auch der Minoritätsladungsträger ab. Die Funktion des Feldeffekttransistors hingegen hängt lediglich vom Fluß der Majoritätsladungsträger ab.
b) Der Feldeffekttransistor hat gegenüber dem Bipolartransistor viele Vorteile. Er ist leichter herzustellen und beansprucht weniger Fläche in einer integrierten Schaltung. Weiterhin rauscht er weniger und besitzt eine höhere Eingangsimpedanz.
c) Der n-Kanal FET wird aus einem Balken n-dotierten Siliziums mit ohmschen Kontakten an den Enden konstruiert. Diese Kontakte an den Enden der Siliziumscheibe werden Source und Drain genannt.
d) Der Fluß von Majoritätsträgern von Source nach Drain kann durch Anlegen einer Potentialdifferenz zwischen Gate und Source gesteuert werden. Das Gate verhält sich also wie eine Steuerelektrode.
e) Wird eine negative Gate-Spannung angelegt, werden positive Ladungen im Kanal induziert, und dies verursacht eine Verarmung an Majoritätsträgern. Dies ist der Grund für die Verwendung des Begriffs Verarmungs-MOSFET.

15 Oscillators and signal generators

A large variety of circuits and instruments have been developed for the generation of periodic and nonperiodic signals. Different types and shapes of signals are required for different purposes. For example the carrier waveform generated by the quartz oscillator in a TV transmitter is very different from the waveform produced by the time-base generator in a TV receiver.

Positive feedback

The factor that is common to all oscillator circuits is positive feedback. It has been seen in chapter 11 that the gain of an amplifier with feedback is given by the expression

$$A_f = \frac{A}{1-\beta A} \qquad \text{where } \beta \text{ is the feedback factor.}$$

With negative feedback the denominator is always greater than one. With positive feedback it is possible to have the condition

$$1-\beta A = 0$$

This gives an infinite value for A_f and implies that the amplifier produces a signal at the output without any signal at the input. This is the condition for oscillation and is known as the Barkhausen criterion. To ensure that oscillation occurs at a single controllable frequency, it is necessary to include a frequency selective network in the feedback path.

Low frequency sine wave oscillators

At low frequencies RC circuits are more commonly used than LC circuits. The reason for this is that at low frequencies, the inductances required for LC circuits are too large and expensive. An RC circuit that is very popular as a low frequency oscillator is the Wien bridge circuit shown in Fig 15.1. This circuit uses an operational amplifier and a frequency selective network (Wien network) composed of R_1C_1 and R_2C_2 . Positive feedback is applied through the Wien network, and negative feedback is applied through R_3 and the bulb. It can be shown that such a circuit will oscillate at a frequency given by the expression

$$f = \frac{1}{2\pi RC}$$

where $C_1 = C_2 = C$ and $R_1 = R_2 = R$

The inclusion of the bulb in the negative feedback circuit tends to stabilize the output of the oscillator. If the output voltage increases, a larger current flows through the bulb and the resistance of its filament increases. This increase in resistance increases the negative feedback, which in turn reduces the output. Even better stabilization of the output can be achieved by replacing the bulb with a fixed resistor and R_3 by a thermistor.

By using ganged (or coupled) variable resistors for R_1 and R_2, an oscillator whose frequency is variable within a limited range can be constructed. Several ranges of frequency can be obtained by using several different pairs of ganged capacitors for C_1 and C_2.

LC oscillators

At frequencies above 50 kHz, the LC resonant circuit is the most commonly used frequency selective network. The circuit of a simple LC oscillator which uses an FET amplifier is shown in Fig 15.2. The primary of the transformer is tuned by a variable capacitor C_2 and its frequency of oscillation can be varied by varying C_2. Positive feedback takes place through the transformer, and the secondary terminals must be connected the correct way round to ensure that the feedback is positive and not negative. If the circuit does not oscillate the secondary connections should be reversed.

Quartz oscillators

Certain crystals, notably quartz and some ceramics exhibit piezoelectric properties. When a crystal with piezoelectric properties is stressed mechanically, a voltage is produced between the two opposite faces of the crystal. Conversely, the application of a voltage across the two faces causes stress and deformation in the crystal. As a result of this property, a piezoelectric crystal vibrates mechanically when an A.C voltage is applied across its faces. The crystal behaves like a resonant circuit of very high Q-factor and low damping. It can be used as the frequency selective network of an extremely stable fixed frequency oscillator. A crystal oscillator circuit using an operational amplifier is shown in Fig 15.3.

Waveform generators

In addition to sine wave oscillators, there are oscillators which can generate other types of waveforms. Among these are pulse generators which generate periodic and single pulse waveforms, triangular waveform generators, and time base generators which produce saw tooth waveforms.

Signal generators

A signal generator is an instrument which can produce a waveform of the desired shape and frequency, and whose output voltage or power can also be set to a definite value. A signal generator uses a suitable oscillator to generate the required type of signal. The output voltage is usually controlled by using a monitoring meter and an attenuator which has a fixed impedance at all settings.

Vocabulary

attenuator	Dämpfungsglied *n*	**ganged capacitor**	Mehrfachdrehkondensator *m*
ceramic	Keramik *f*	**generate**	erzeugen *v*
composed of	bestehen aus *v*	**monitoring meter**	Überwachungsmeßgerät *n*
controllable	kontrollierbar, regulierbar *v*	**monostable oscillator**	monostable Kippschaltung *f*
conversely	umgekehrt *adv*	**oscillator**	Oszillator *m*
criterion	Kriterium *n*	**piezoelectric**	piezoelektrisch *adj*
damping	Dämpfung *f*	**popular**	beliebt *adj*
definite value	bestimmter Wert *m*	**require**	brauchen, erfordern *v*
denominator	Nenner *m*	**shape**	Form *f*
different	anders, verschieden *adj*	**signal generator**	Meßsender *m*
desired shape	gewünschte Form *f*	**thermistor**	Heißleiter *m*
develop	entwickeln *v*	**timebase generator**	Zeitablenkung *f*
expensive	teuer *adj*	**variable capacitor**	Drehkondensator *m*
exhibit	ausstellen, aufweisen *v*	**vibrate**	schwingen *v*
frequency selective network	frequenzselektive Schaltung *f*	**waveform generator**	Wellenformgenerator *m*

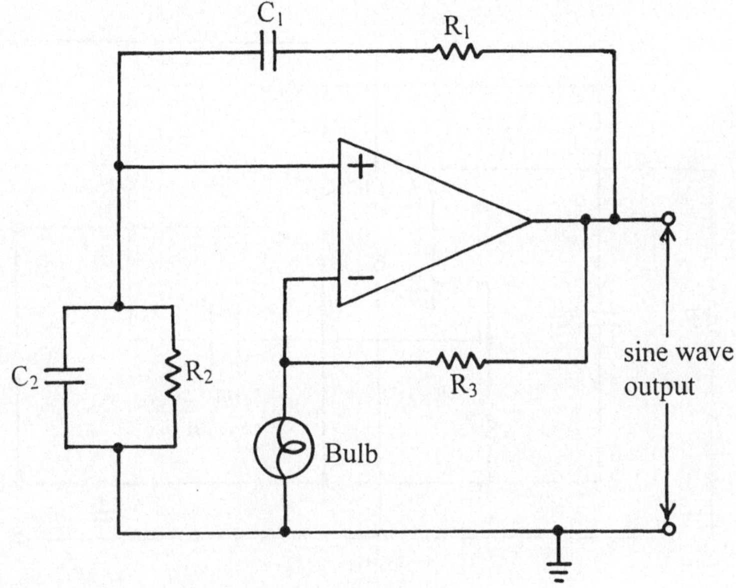

Fig 15.1 Wien bridge oscillator

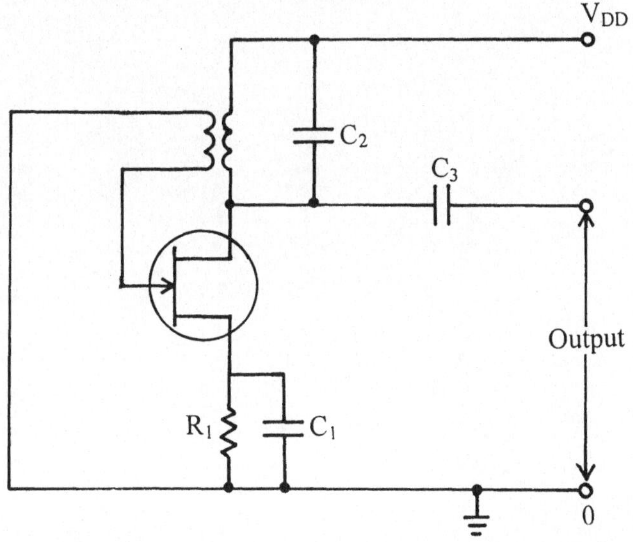

Fig 15.2 LC oscillator using a FET

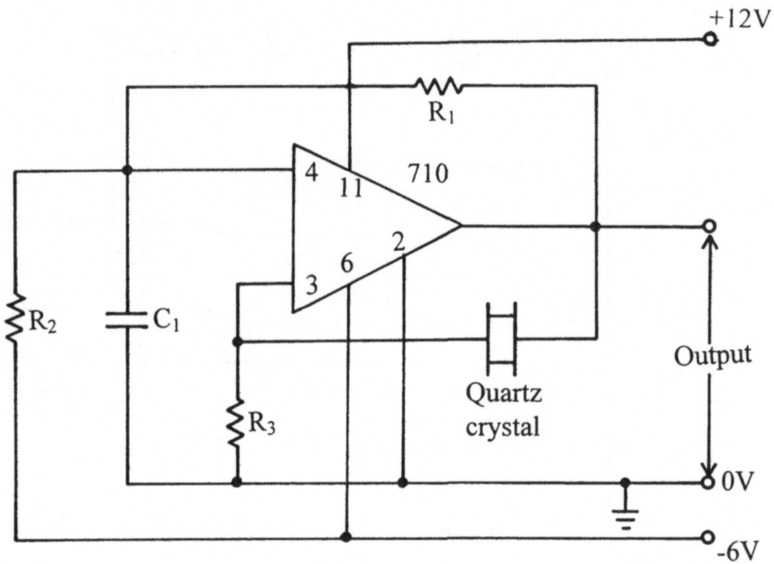

Fig 15.3 Quartz crystal oscillator with a rectangular wave output

Exercises XV

1. Answer the following questions:

a) What factor is common to all oscillator circuits?
b) What happens when the gain of an amplifier with feedback has an infinite value ?
c) What has to be done to ensure that oscillation takes place at a single controllable frequency ?
d) What is the condition for oscillation ?
e) What types of oscillator circuits are commonly used at low frequencies ?
f) Give the name of a circuit that is very popular as the basis for the construction of a low frequency oscillator.
g) At what frequencies are LC oscillators used ?
h) What happens when a piezoelectric crystal is mechanically stressed ?
i) Why does a quartz crystal oscillator have a high frequency stability ?
j) How is the output level of a signal generator controlled ?

2. Fill in the gaps in the following sentences:

a) The ____ that is common to all oscillator circuits is ____.
b) To ensure that oscillation occurs at a ____ it is necessary that a ____ be included in the feedback path.
c) At low frequencies, ____ are more commonly used than ____.
d) The inclusion of the bulb in the ____ tends to ____ the output of the oscillator.
e) At frequencies above 50 kHz, the ____ circuit is the most commonly used ____ network.
f) Certain crystals, notably ____ and some ceramics, exhibit ____ properties.
g) When a piezoelectric crystal is ____ mechanically, a ____ is produced between two opposite faces of the crystal.
h) A piezoelectric crystal ____ mechanically when an ____ is applied across two of its faces.
i) A ____ can be used as the frequency selective network of an ____ fixed frequency oscillator.

j) The carrier waveform produced by a ____ in a TV transmitter is very different from the waveform produced by the ____ generator in a TV receiver.

3. Translate into English:

a) Eine Vielzahl verschiedener Schaltungen ist zur Erzeugung von periodischen und nicht-periodischen Signalen entwickelt worden, da unterschiedliche Signalarten und -formen für verschiedene Anwendungen benötigt werden.
b) Die Trägerwellenform, die vom Quartzoszillator in einem Fernsehsender erzeugt wird, unterscheidet sich stark von der Wellenform, die von der Zeitablenkung eines Fernsehempfängers erzeugt wird.
c) Bestimmte Kristalle, insbesondere Quartz und einige Keramiken, weisen piezoelektrische Eigenschaften auf. Wird der Kristall mechanisch beansprucht, so entsteht eine elektrische Spannung zwischen den entgegengesetzten Stirnflächen des Kristalls.
d) Der Kristall verhält sich wie ein Schwingkreis mit sehr hohem Q-Faktor und geringer Dämpfung. Er kann als Frequenzfilter in einem extrem stabilen Festfrequenzoszillator eingesetzt werden.
e) Ein Signalgenerator ist ein Instrument, welches eine Wellenform gewünschter Form und Frequenz erzeugen kann und dessen Ausgangsspannung oder -leistung ebenfalls auf einen festen Wert eingestellt werden kann.

16 Silicon controlled rectifiers or thyristors

Solid state devices play an important part in electrical power control, particularly in applications which require a variable but controlled amount of current. Among these are motor speed control, electrical welding, and lighting control. The use of silicon controlled rectifiers has made electrical power control an efficient and inexpensive process.

The four layer diode

The four layer diode consists of four layers of silicon arranged in the order p-n-p-n as shown in Fig 16.1. When the anode is made positive with respect to the cathode (forward bias), the junctions J_1 and J_3 are forward biased while J_2 is reverse biased. With forward bias, the diode has two stable states, one a high resistance state called the OFF state, and the other a low resistance state called the ON state.

The volt-ampere characteristic of a four layer diode

The volt-ampere characteristic of a four layer diode is shown in Fig 16.2. If a forward voltage is applied, only a small current flows until the voltage reaches the firing or break-over voltage V_{B0}. The current corresponding to V_{B0} is I_{B0}. If the forward voltage is increased beyond V_{B0}, the diode switches from its OFF (blocked) state to its ON (saturated) state, and now operates in the saturation region.

If the voltage is reduced, the diode remains in the ON state until the current decreases to I_H. The current I_H and the corresponding voltage V_H, are called the holding or latching current and voltage respectively. If the current decreases below I_H, the diode switches to the OFF state.

The silicon controlled rectifier or thyristor

As its name implies, the silicon controlled rectifier (SCR) is a rectifier which has the added ability to control the power delivered to the load. The SCR has a structure similar to the four layer diode with an additional electrode called the gate as shown in Fig 16.3. The gate acts as a control electrode, and currents flowing in the gate-cathode circuit may be used to control the anode to cathode break-over voltage.

The volt-ampere characteristics of the SCR

The volt-ampere characteristics of an SCR are shown in Fig 16.4 for different values of cathode-gate currents. The break-over voltage will be seen to be a function of the gate current. The break-over or firing voltage decreases with increasing gate current.

Simple power control circuit

A simple circuit which can be used to control the power supplied to a load is shown in Fig 16.5. An A.C voltage is applied through a load resistor between the anode and cathode of an SCR. The peak value of the applied voltage is less than the break-over voltage corresponding to zero gate current.

The rectifier remains OFF and no current flows through it until it is turned ON by the application of a narrow gate current pulse, which lowers the break-over voltage to a value less than the peak value of the applied A.C voltage. The rectifier conducts for only part of the cycle, and the conduction angle is controlled by the phase of the gate pulses relative to the A.C supply. Fig 16.6 shows the relationship between the output voltage and the gate pulses. The power supplied to the load can be varied from zero to a maximum value by varying the phase of the gate pulses.

The triac or bidirectional thyristor

The SCR, although ideally suited for A.C power control, has one serious disadvantage. It is a half-wave device and is therefore only able to supply half power even at full conduction. This disadvantage may be overcome by connecting two thyristors in inverse-parallel to provide full-wave operation. This arrangement however requires two isolated but synchronized sources of gate pulses.

The most useful device available for A.C power control is the triac or bidirectional thyristor. The construction of this device is shown in Fig 16.7 and it can be considered to be composed of two inverse-paralleled thyristors which are both controlled by a single gate electrode. The terms anode and cathode have no meaning in a triac. The contact next to the gate is called main terminal 1 (MT1) and the other is called main terminal 2 (MT2).

A triac application

A simple triac application circuit is shown in Fig 16.8. This circuit can be used for example as a motor speed control circuit or a lamp dimmer circuit.

The gate trigger pulses are produced by a diac or bidirectional trigger diode. It has a low avalanche breakdown voltage of about 30 V. When the voltage across the capacitor reaches the breakdown voltage of the diac, a sudden pulse from the capacitor triggers the triac. The phase of the pulses and hence the power supplied, can be controlled by varying the resistor R_2. The voltage across the load corresponding to different phase angles is shown in Fig 16.9.

Vocabulary

arrangement	Anordnung *f*	**power control**	Leistungsregelung *f*
bidirectional	in zwei Richtungen *adj*	**pulse**	Puls *m*
blocked state	Sperrzustand *m*	**relation**	Beziehung *f*
break-over voltage	Kippspannung *f*	**saturated**	gesättigt *adj*
conduction angle	Phasenanschnittswinkel *m*	**silicon controlled rectifier**	regelbarer Siliziumgleichrichter, Thyristor *m*
diac	Diac *m*	**speed control**	Drehzahlregelung *f*
efficient	leistungsfähig, wirksam *adj*	**stable state**	Ruhezustand *m*
firing voltage	Zündspannung *f*	**synchronize**	synchronisieren *v*
imply	bedeuten *v*	**therefore**	deshalb *adv*
inexpensive	nicht teuer *adj*	**thyristor**	Thyristor *m*
inverse-parallel	umgekehrt-parallel *adj*	**triac**	Triac, Zweirichtungsthyristor *m*
latch	selbsthaltender Schalter *m*	**various**	mehrere *adj*
peak value	Scheitelwert *m*	**weld**	schweißen *v*

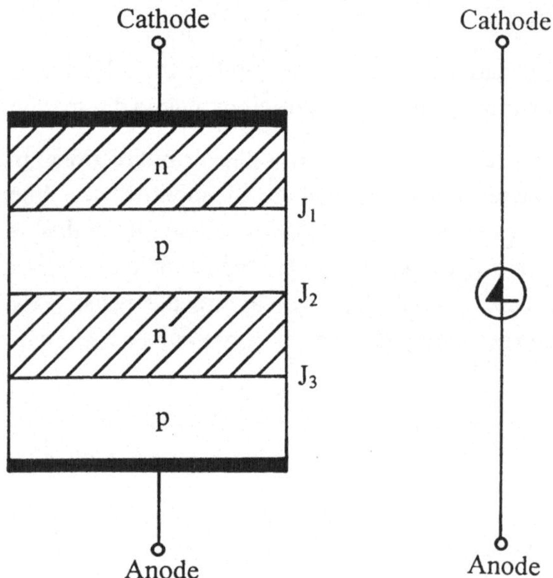

Fig 16.1 Construction of a four layer diode

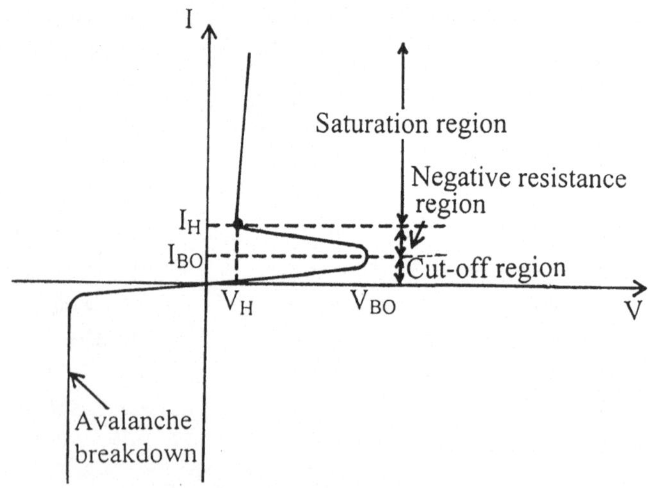

Fig 16.2 Volt-ampere characteristic of a four layer diode

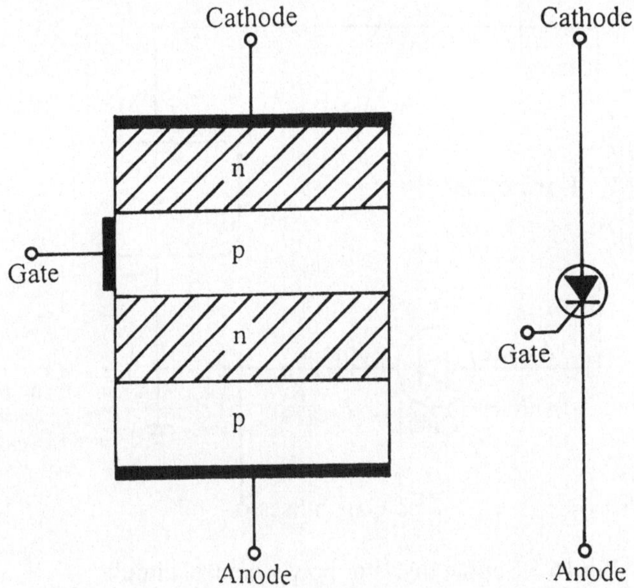

Fig 16.3 Construction and circuit symbol of a SCR

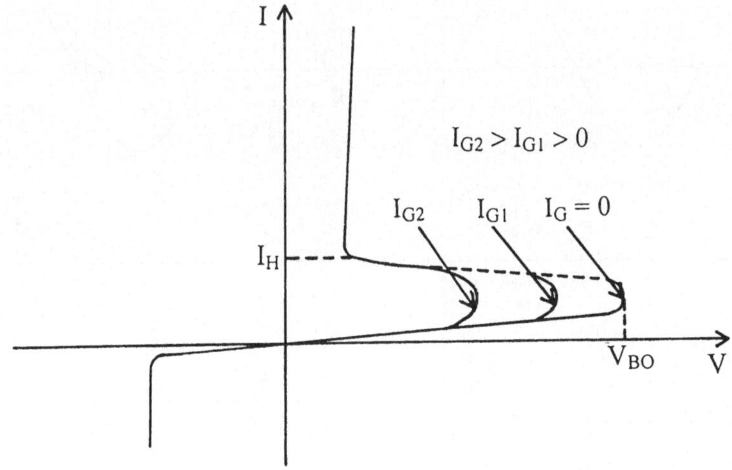

Fig 16.4 Volt-ampere characteristic for a SCR

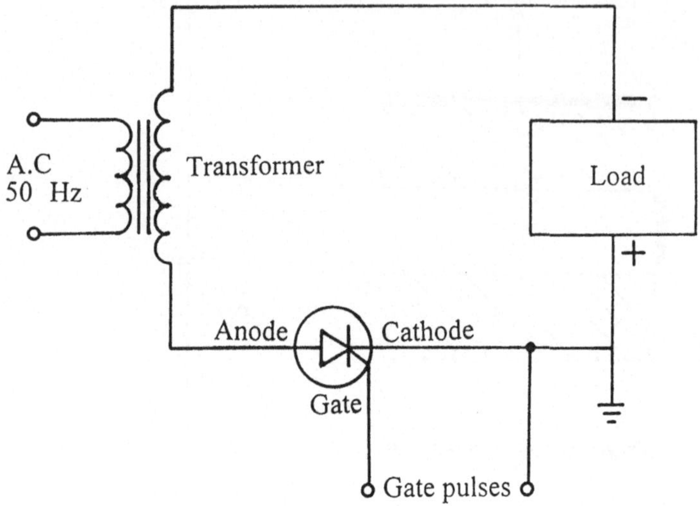

Fig 16.5 Simple thyristor power control circuit

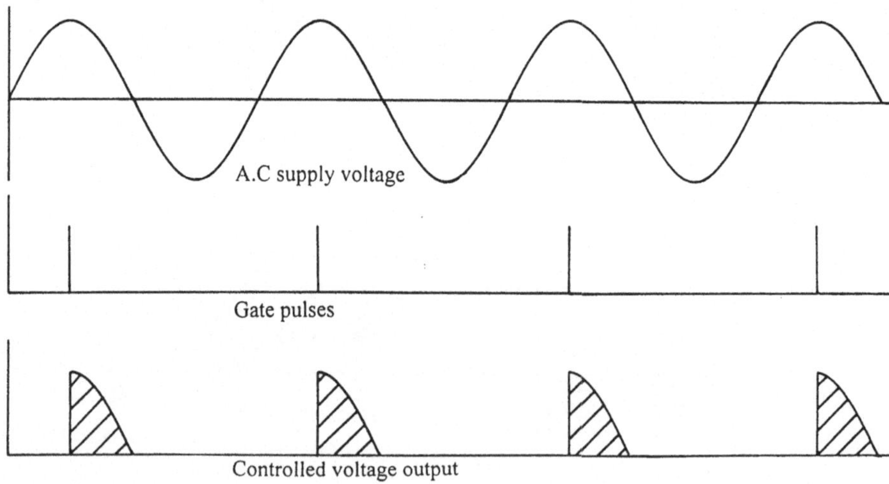

Fig 16.6 Gate pulses and controlled voltage output of a simple thyristor circuit

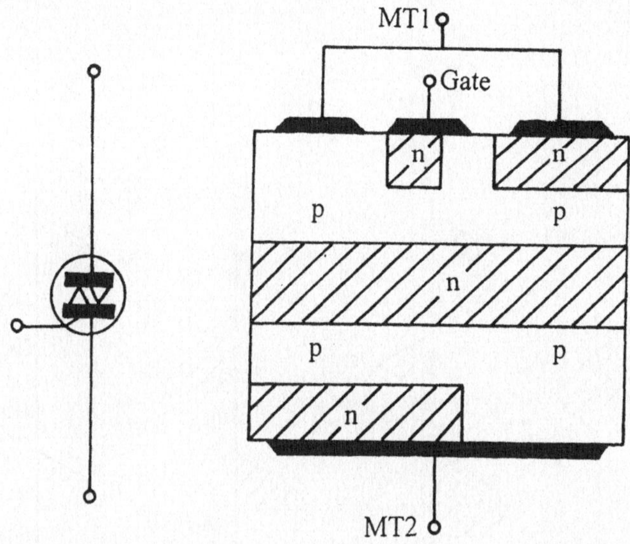

Fig 16.7 The construction of a triac and its circuit symbol

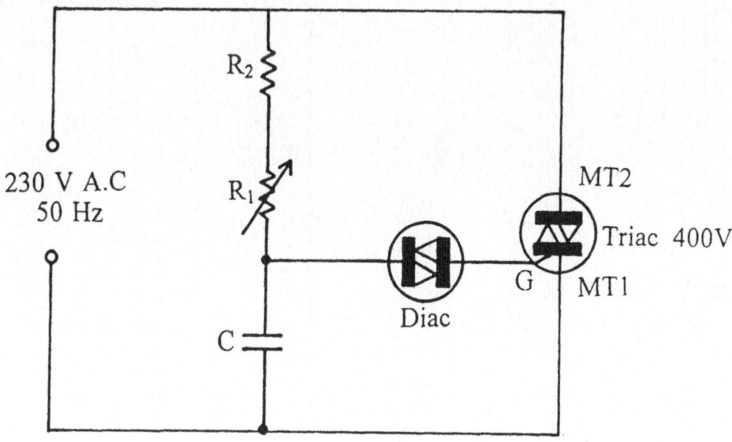

Fig 16.8 Simple triac power control circuit

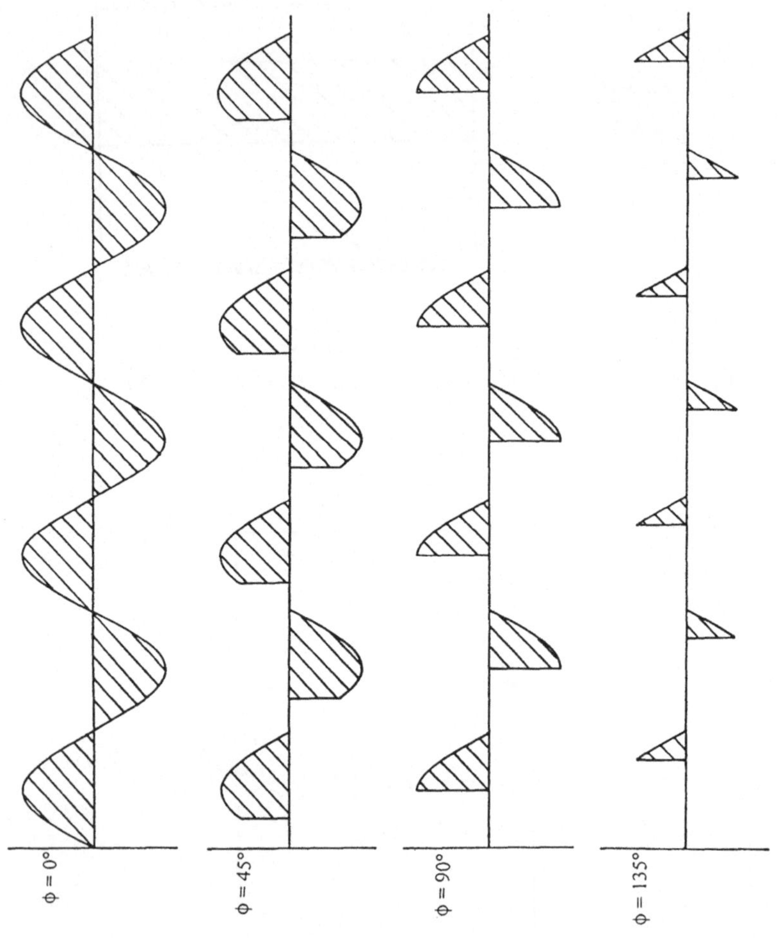

Fig 16.9 Voltage waveforms across the load of a triac for different values of phase angle

Exercises XVI

1. Answer the following questions:

a) What advantages do silicon controlled rectifiers have over ordinary rectifiers in electrical power control ?
b) Give some examples of power control applications in which a SCR may be used to advantage.
c) How are the layers of semiconductor material in a four layer diode arranged ?
d) What are the stable states of a four layer diode called ?
e) What happens when the applied voltage exceeds the breakover voltage V_{BO} ?
f) What is meant by the latching voltage and the latching current ?
g) What is the function of the gate in an SCR, and what do the gate-cathode currents control ?
h) How can two SCR devices be used to deliver full power ?
i) What advantage does the triac have over the SCR ?
j) How can the conduction angle in a thyristor be varied ?

2. Fill in the gaps in the following sentences:

a) The use of silicon controlled rectifiers has made electrical ____ an efficient and ____ process.
b) A four layer diode is composed of ____ arranged in the order ____.
c) A ____ four layer diode has two ____ states.
d) If the ____ is increased beyond V_{BO}, the diode ____ from its OFF state to its ON state.
e) The SCR is a ____ which has the added ability to control the ____.
f) The gate acts as a ____ and currents flowing in the gate-cathode circuit may be used to control the anode to cathode ____.
g) The SCR remains OFF and no current ____ until it is turned ON by the application of a ____.
h) The SCR is a half-wave device and is only able to ____ even at full ____.
i) The triac may be considered to be composed of two ____ thyristors which are both controlled by a ____.

j) The ____ angle is controlled by the ____ relative to the A.C supply.

3. Translate into English:

a) Festkörperbauelemente spielen heute eine wichtige Rolle in der elektrischen Leistungsregelung. Viele Anwendungen benötigen einen variablen aber geregelten Strom.
b) Unter diesen Anwendungen sind Motordrehzahlregelung, elektrisches Schweißen und Beleuchtungsregelung. Der Einsatz von regelbaren Siliziumgleichrichtern hat den Prozess der elektrischen Leistungsregelung preiswert und leistungsfähig gemacht.
c) Wird die Anode eines Thyristors in Bezug auf die Kathode positiv, so sind die Übergänge J_1 und J_3 in Durchflußrichtung vorgespannt, während J_2 in Sperrichtung vorgespannt ist. Bei Vorspannung in Durchflußrichtung besitzt die Diode zwei stabile Zustände, einen hochresistiven OFF genannten und einen sehr niederresistiven als ON bezeichneten Zustand.
d) Wie der Name nahelegt, handelt es sich bei dem regelbaren Siliziumgleichrichter um einen Gleichrichter, der zusätzlich über die Fähigkeit verfügt, die an die Last abgegebene Leistung regeln zu können. Der regelbare Siliziumgleichrichter besitzt eine ähnliche Struktur wie die Vierschichtdiode, mit einer zusätzlichen als Gate bezeichneten Elektrode, wie in den Abbildungen dargestellt.
e) Das nützlichste Bauelement bei der Wechselstromleistungsregelung ist der Triac oder bidirektionale Thyristor. Dieser kann als aus zwei invers parallel liegenden Thyristoren bestehend aufgefaßt werden, die beide von einem gemeinsamen Gate kontrolliert werden.

17 Logic circuits

Logic circuits form the basis of a large number of electronic systems like computers, data processing systems, and digital communication systems. Digital systems possess many advantages over analog systems and there is an increasing trend towards converting analog signals into digital signals and then using digital systems to process the signals further.

A digital system uses binary devices which can have only two states. Various names have been given to these states like Low and High or 0 and 1. An example of an electronic device which can operate in such a fashion is a transistor which is allowed to be in the cut-off region or in the saturation region, but not in the intermediate active region.

A digital system however complicated, functions by the repetition of a few basic operations. The circuits used to perform these operations are called logic gates. The simplest of the logic gates are called OR, AND, and NOT. These are the building blocks from which more complicated digital systems are built. The type of algebra that is relevant to digital systems is called Boolean algebra, and logic gates are used to implement Boolean algebraical expressions.

The OR gate

The symbol and truth table for an OR gate are shown in Fig 17.1. An OR gate has two or more inputs and one output. Its operation is in accordance with the definition:

The output of an OR gate is in the 1 state, if one or more inputs are in the 1 state.

The Boolean expression for this gate is

$$Y = A + B + \ldots\ldots + N$$

and is read "Y equals A or B or ..... or N".

The AND gate

The symbol and truth table for an AND gate are shown in Fig 17.2. An AND gate has two or more inputs and one output. Its operation is in accordance with the definition:

The output of an AND gate is in the 1 state, if and only if all the inputs are in the 1 state.

The Boolean expression for this gate is given by

$$Y = A\,B \,.....\, N$$

and is read "Y equals A and B and ..... and N".

The NOT gate

The symbol and truth table for a NOT gate are shown in Fig 17.3. A NOT gate has only one input and one output. The operation of a NOT gate is in accordance with the definition:

The output of a NOT gate is in the 1 state, only if the input is not in the 1 state.

The Boolean expression for this gate is given by

$$Y = \overline{A}$$

and is read as "Y equals not A", or "Y is the complement of A".

The NOR gate

The NOR gate can be considered to be a combination of an OR gate and a NOT gate as shown in Fig 17.4. It operates in accordance with the definition:

The output of a NOR gate is in the 1 state, if and only if all inputs are in the 0 state.

The Boolean expression for the NOR gate is given by

$$Y = \overline{A + B + \;....\; + N}$$

and is read as "Y equals not A or B or ..... or N".

The NAND gate

A NAND gate can be considered to be a combination of an AND gate and a NOT gate as shown in Fig 17.5. It operates in accordance with the definition:

The output of a NAND gate is in the 1 state, if one or more inputs are in the 0 state. The Boolean expression for a NAND gate is given by

$$Y = \overline{A\,B\,C\,....\,N}$$

and is read as "Y equals not A and B and ..... and N".

The EXCLUSIVE OR (XOR) gate

The OR gate recognizes the presence of one or more 1s (ones) at the input.

The EXCLUSIVE OR gate recognizes only an odd number of 1s at the input. The operation of a two input EXCLUSIVE OR gate is in accordance with the definition:

The output of a two input EXCLUSIVE OR gate is in the 1 state, if and only if one input is in the 1 state.

The standard symbol and the truth table are shown in Fig 17.6. The Boolean expression for this gate is given by

$$Y = (A + B)(\overline{AB})$$

The gate may be constructed from simpler gates in many ways. One way of constructing this gate is shown in Fig 17.7. The EXCLUSIVE OR gate can be used as an equality detector.

De Morgan's laws

De Morgan's laws show that certain types of circuits are logically equivalent.

There are two laws, and these are best expressed in terms of the following Boolean equations:

$$\overline{A\,B\,C\,.....\,N} = \overline{A} + \overline{B} + \overline{C} + ...+\overline{N}$$

$$\overline{A + B + C + \,.....\, + N} = \overline{A}\,\overline{B}\,\overline{C}\,.....\,\overline{N}$$

Using De Morgan's laws, the complement of a Boolean function can be found by changing all OR operations to AND operations, all AND operations to OR operations, and then negating each binary symbol.

Fig 17.8(a) shows how an OR gate can be converted into an AND gate, by inverting all the inputs and also the output. Fig17.8(b) shows how an AND gate can be converted into an OR by complementing all inputs and the output.

Vocabulary

basis	Fundament *n*, Basis *f*	**equality detector**	Äquivalenzschaltung *f*
binary device	Binärbauelement *n*	**fashion**	Mode *f*
Boolean algebra	Boolesche Algebra *f*	**implement**	ausführen, in Kraft setzen *v*
combination	Verknüpfung, Kombination *f*	**in accordance with**	in Übereinstimmung mit *adv*
complicated	kompliziert *adj*	**logic gate**	logisches Gatter *n*
complement	Komplement *n*, Ergänzung *f*	**logically equivalent**	logisch gleichwertig *adj*
convert	umwandeln *v*	**possess**	besitzen *v*
data processing system	Datenverarbeitungssystem *n*	**recognize**	erkennen *v*
digital circuit	Digitalschaltung *f*	**repetition**	Wiederholung *f*
digital communication system	digitales Nachrichtensystem *n*	**trend**	Richtung *f*,Trend *m*
		truth table	Wahrheitstabelle *f*

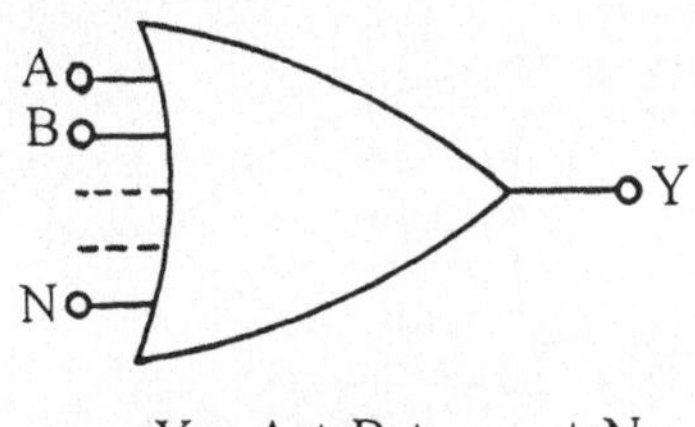

$Y = A + B + ...+ N$

Input		Output
A	B	Y
0	0	0
0	1	1
1	0	1
1	1	1

Fig 17.1 Symbol and truth table for an OR gate

$Y = A\,B\,.........N$

Input		Output
A	B	Y
0	0	0
0	1	0
1	0	0
1	1	1

Fig 17.2 Symbol and truth table for an AND gate

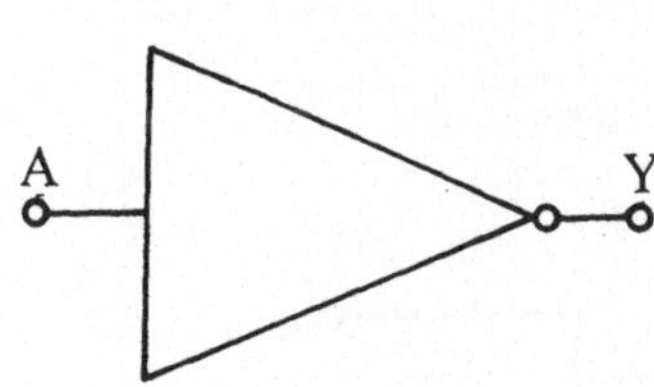

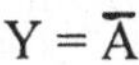
$Y = \overline{A}$

Input	Output
A	Y
0	1
1	0

Fig 17.3 Symbol and truth table for an NOT gate

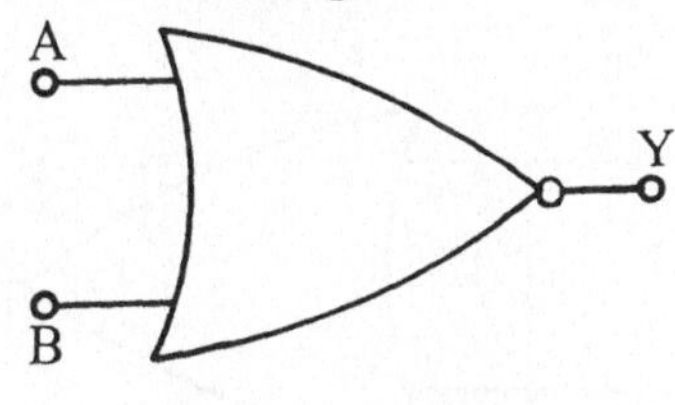

$Y = \overline{A + B}$

Input		Output
A	B	Y
0	0	1
0	1	0
1	0	0
1	1	0

Fig 17.4 Symbol and truth table for an NOR gate

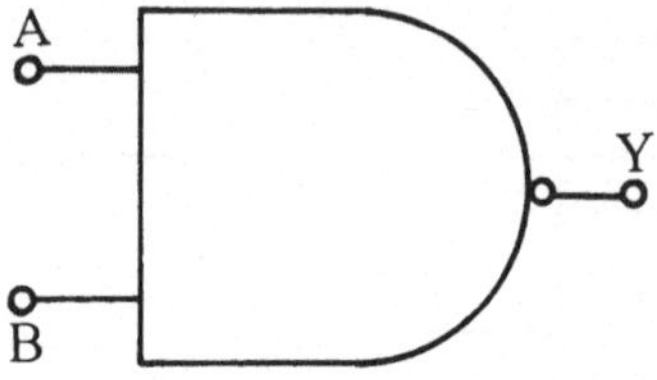

Input		Output
A	B	Y
0	0	1
0	1	1
1	0	1
1	1	0

Fig 17.5 Symbol and truth table for an NAND gate

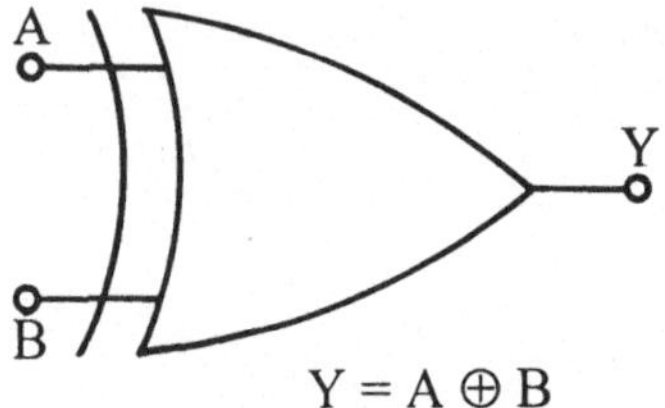

Input		Output
A	B	Y
0	0	0
0	1	1
1	0	1
1	1	0

Fig 17.6 Symbol and truth table for an EXCLUSIVE OR gate

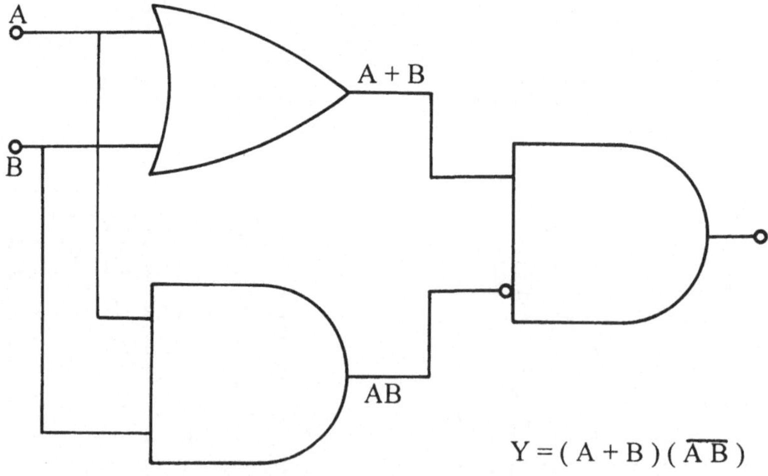

Fig 17.7 A logic block diagram for an EXCLUSIVE OR gate

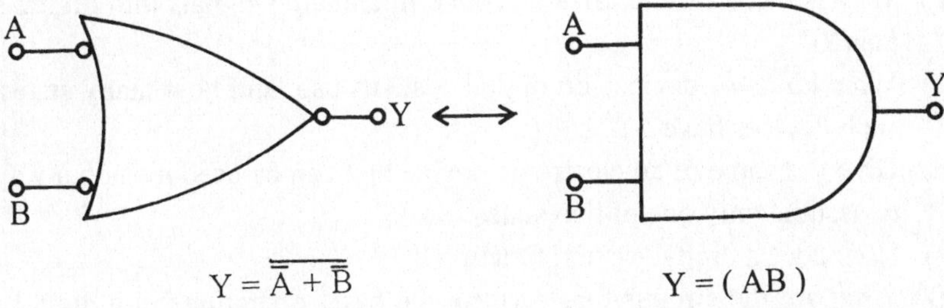

Fig 17.8 (a) Conversion of an OR gate into an AND gate

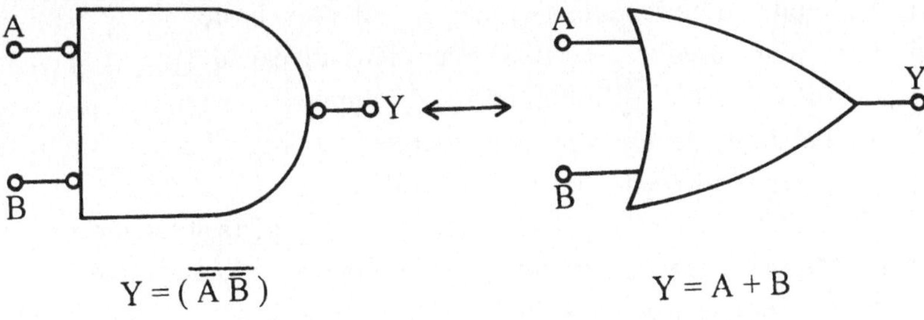

Fig 17.8 (b) Conversion of an AND gate into an OR gate

Exercises XVII

1. Answer the following questions:

a) Give the names of some systems which are based on digital circuits.
b) Why is there a trend towards converting analog signals into digital signals ?
c) What kinds of devices do digital systems use, and how many states do such devices have ?
d) Give the name of an electronic device that can be used in such a way as to assume only one of two states.
e) How does a digital system function ?
f) What are the circuits that perform the basic operations in a digital circuit called ?
g) What is the difference between a two input OR gate and a two input EXCLUSIVE OR gate ?
h) Define the operation of an AND gate.
i) How can an EXCLUSIVE OR gate be constructed ?
j) How can an OR gate be converted into an AND gate ?

2. Fill in the blanks in the following sentences:

a) Digital circuits form the basis of a large number of ____ systems, like computers and ____ processing systems.
b) A digital system uses ____ which can only have ____ states.
c) A digital system functions by the ____ of a few basic ____.
d) The circuits used to ____ these operations are called ____.
e) An OR gate has ____ inputs and ____ output.
f) An electronic device which can operate in a binary fashion is a transistor which is allowed to be at ____ or at ____.
g) The EXCLUSIVE OR gate ____ only an ____ of 1s at the input.
h) De Morgan's laws show that ____ circuits are ____ equivalent.
i) An OR gate can be converted into an ____gate by ____ all the inputs and also the output.
j) The type of algebra that is ____ to digital systems is called ____ .

3.Translate into English:

a) Digitalschaltungen stellen die Grundlage einer Vielzahl von elektronischen Systemen dar, wie Computer, Datenverarbeitungssysteme und digitale Nachrichtensysteme.

b) Ein digitales System, wie kompliziert es auch sei, funktioniert durch die Wiederholung weniger grundlegender Operationen. Die zur Ausführung dieser Operationen eingesetzten Schaltungen werden als logische Gatter bezeichnet.

c) Der Ausgang eines Exklusiv-ODER-Gatters mit zwei Eingängen befindet sich im 1 Zustand, falls sich nur einer der Eingänge im 1 Zustand befindet. Das Gatter kann auf verschiedene Weisen aus einfacheren Gattern zusammengesetzt werden. Eine Möglichkeit dieses zu erreichen ist in Abb. 17.7 dargestellt.

d) Diese Gatter sind die Bausteine, aus denen kompliziertere digitale Systeme aufgebaut werden. Die für digitale Systeme relevante Algebra wird Boolesche Algebra genannt.

e) Ein ODER-Gatter verfügt über zwei oder mehr Eingänge und einen Ausgang. Seine Funktionsweise folgt der Definition: der Ausgang eines Oder-Gatters befindet sich im 1 Zustand, wenn einer oder mehrere Eingänge sich ebenfalls im 1 Zustand befinden.

18 Logic families

Many types of electronic circuits have been used to implement logic gates. These are almost invariably manufactured in the form of integrated circuits. The integrated circuits used are classified into logic families depending on the type of circuits used for the gates. Among the families are:

- DTL Diode Transistor Logic
- TTL Transistor Transistor Logic
- ECL Emitter Coupled Logic
- CMOS Complementary MOS Logic

Some of the factors which have to be considered in selecting a logic family for a particular application are the following:

- Power dissipation
- Speed
- Cost
- Fanout
- Availability

It is desirable that the power dissipation be as small as possible. CMOS gates have the lowest power dissipation, and they are therefore very attractive from this point of view. The speed of a logic family is measured in terms of the propagation delay time of its basic NAND gate. ECL is the best choice for high speed circuits. Fanout is defined as the number of loads that can be driven from a single source. The term availability refers to two things, the popularity of the logic family, and the breadth of the family, meaning the number of different types of gates and circuits commerciably available in the family.

TTL Logic

Standard TTL chips are available in a larger variety of circuits than other families. They are faster than CMOS but slower than ECL. Their output impedance is low in both the 0 and 1 states.

The 7400 TTL IC series is extremely popular. In addition to the standard TTL family of chips, other TTL families like high speed TTL, low power TTL, and Schottky TTL are also available.

The NAND gate

The backbone of the 7400 series is the multiple emitter NAND gate. A circuit diagram of the gate is shown in Fig 18.1, and a brief description of its operation is given below.

The input transistor T_1 has several emitters, and each input is connected to one of the emitters. The output from T_1 drives T_2 which acts as a phase-splitter circuit. Transistor T_2 in turn drives T_3 and T_4 which are called totem pole transistors.

When all inputs are 1, all the emitter-base junctions of T_1 are reverse biased. A current flows through T_2 and saturates it. This produces a low voltage on the base of T_3 cutting it off. The current flowing through T_2 enters the base of T_4 and saturates it. Since T_4 is on and T_3 is off, the output is at a low voltage or in the 0 state. Therefore, when all inputs are 1 the output is 0 which corresponds to the action of a NAND gate.

If one or more inputs are low, current flows through one of the multiple emitter inputs, and T_2 is cut-off because of a lack of base current. T_4 also receives no base current and remains off. T_3 is however on because there is a high voltage on the collector of T_2 and therefore on the base of T_3. The output is connected through T_3 to a high voltage. Therefore when any of the inputs are are 0, the output is 1, and this confirms the action of a NAND gate.

Standard TTL chips

Most of the circuits in the 7400 TTL series are small scale integrated (SSI) circuits. As an example, consider the number 7400 IC which is a specific chip in the series. It consists of a quad (four independent) 2-input NAND gates, in a 14 pin package as shown in Fig 18.2. Pin 7 is connected to ground and pin 14 is reserved for V_{CC}. Most of the SSI chips are in 14 pin packages, and the number of independent circuits in each package depend on the number of pins required for each circuit. Details of some of the chips in the 7400 series are given below:

- 7404 Hex inverters (six independent inverters)
- 7410 Triple 3-input NAND gates (three independent gates)
- 7420 Dual 4-input NAND gates (two independent gates)
- 7486 Quad 2-input XOR gates (four independent EXCL. OR gates)

CMOS chips

The other family of chips that has become extremely popular is the CMOS family. Its popularity is due to the lower power dissipation of the gates and the small space they occupy in an integrated circuit.

A CMOS gate is composed of two enhancement mode MOS gates. The circuit of a CMOS inverter is shown in Fig 18.3. It consists of a n-channel transistor and a p-channel transistor connected in series.

The input voltage is either at 0 V (logic 0) or $-V_{DD}$ (logic 1). When V_i is at $-V_{DD}$ T_2 is on, but draws neglible steady state current. T_1 is turned off and the output V_o is at 0 V. The input therefore appears inverted at the output. When 0 V is applied at the input T_1 is on and T_2 is off. The output is now at $-V_{DD}$ (logic 1) and here again the input has been inverted.

Negligible current is drawn by the transistors in the steady state and therefore power consumption is low. More power is required when switching from one state to another. The power consumed by a CMOS gate is proportional to the frequency at which it is switched.

CMOS gates are slower in action and not available in such a variety of circuits as TTL gates.

The CMOS MM74C00 series

A very useful series of CMOS gates is the MM74C00 series, in which each IC has the same function as the equivalent TTL gate in the 7400 series. In some cases, the CMOS gates can be interfaced directly with the TTL gates.

Vocabulary

action	Wirkung, Handlung *f*	**lack of**	Mangel an *n*
attractive	anziehend, reizvoll *adj*	**multiple**	mehrfach *adj*
availability	Verfügbarkeit *f*	**particular**	besonders, speziell *adj*
classify	klassifizieren, einteilen *v*	**phase-splitter**	Phasenteiler *m*
confirm	bestätigen *adj*	**point of view**	Standpunkt *m*
counter	Zähler *m*	**power dissipation**	Leistungsaufnahme *m*
factor	Faktor *m*, Element *n*	**propagation delay time**	Laufzeitverzögerung *f*
fanout	Mehrfachschnittstelle *f*	**select**	auswählen *v*
interfere	stören, einmischen *v*	**specific**	spezifisch, bestimmt *adj*
invariably	ausnahmslos *adv*	**switch**	schalten *v*
interface	anschließen *v*, Schnittstelle *f*	**unique**	einmalig, einzig *adj*

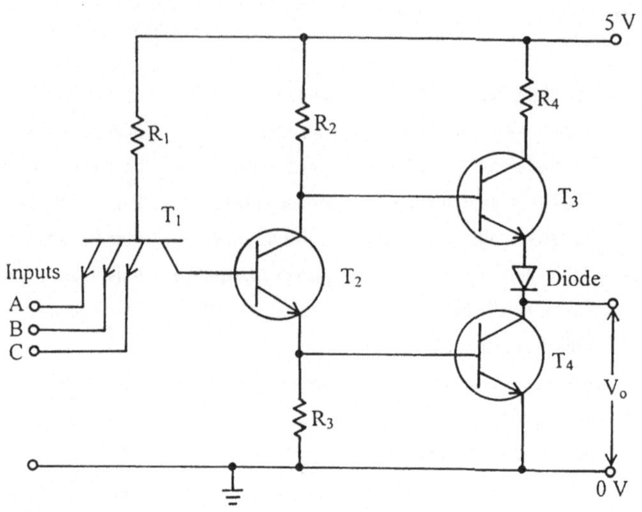

Fig 18.1 A TTL multiple emitter NAND gate

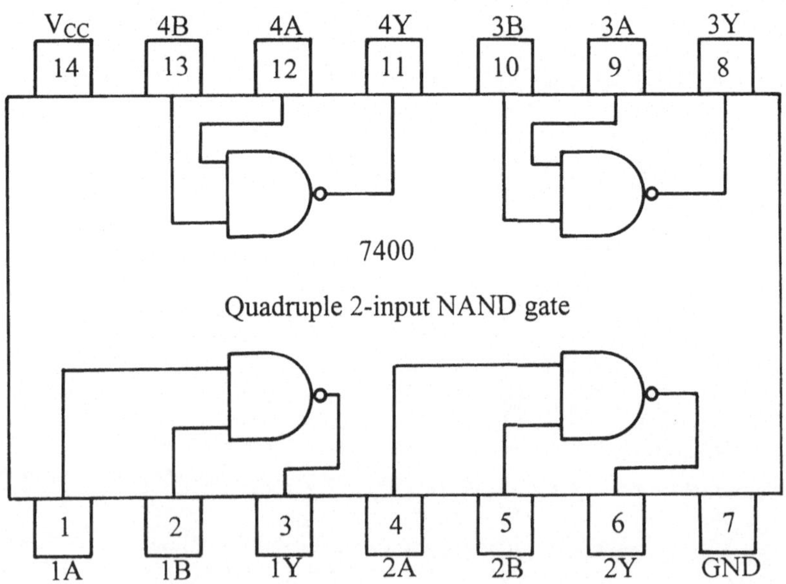

Fig 18.2 Pin layout diagram for the 7400 IC

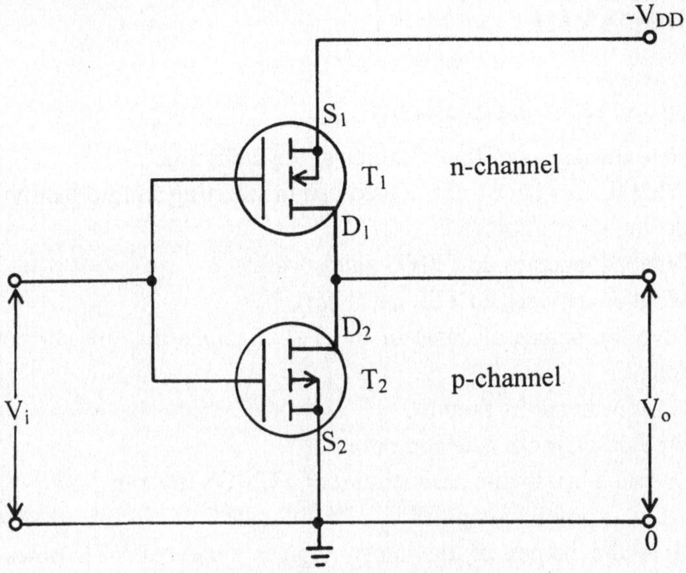

Fig 18.3 A CMOS inverter circuit

Exercises XVIII

1. Answer the following questions:

a) Give the names of four families of logic circuits.
b) What factors need to be considered in selecting a logic family for a particular application ?
c) What advantages do CMOS gates possess ?
d) What advantages do TTL gates have ?
e) Give the names of some of the TTL families that are currently available.
f) What is meant by fanout ?
g) What is a CMOS gate composed of ?
h) Describe briefly the basic circuit of a CMOS inverter.
i) When is power dissipated in a CMOS gate ?
j) Give the names of the most popular series of TTL gates and the equivalent CMOS series.

2. Fill in the gaps in the following sentences:

a) Many types of ____ have been used to ____ logic gates.
b) Fanout is defined as the number of ____ that can be driven from a ____.
c) The integrated circuits are classified into ____ depending on the type of circuits used for the ____.
d) The ____ of a logic family is measured in terms of the ____ time of its basic ____ gate.
e) CMOS gates are popular because of their ____ and the ____ that they occupy in an integrated circuit.
f) Power is dissipated in a ____ gate only when it ____ from one state to another.
g) Standard ____ chips are available in a larger number of circuits than other ____.
h) The power ____ in a CMOS gate is proportional to ____ at which the gate is switched.
i) CMOS gates are ____ in action than TTL gates and are not available in such a large ____ as TTL gates.

j) A useful series of CMOS gates is the ____ series in which each IC has the same function as the ____ TTL in the 7400 gate series.

3. Translate into English:

a) Zur Verwirklichung logischer Gatter sind viele Arten elektronischer Schaltungen eingesetzt worden. Diese Gatter werden ausnahmslos in Form integrierter Schaltungen hergestellt.
b) Standard TTL ICs sind in einer größeren Anzahl an unterschiedlichen Schaltungen erhältlich als andere Logikfamilien. Sie sind schneller als CMOS Schaltungen, aber langsamer als ECL Schaltungen.
c) CMOS ICs sind äußerst beliebt, da sie eine geringe Leistungsaufnahme aufweisen und wegen der kleinen Fläche, die sie in einer integrierten Schaltung einnehmen.
d) In einem CMOS Gatter wird Leistung lediglich verbraucht, wenn es von einem Zustand in den anderen schaltet. Der Leistungverbrauch ist somit proportional zur Schaltfrequenz, mit der das Gatter betrieben wird.
e) Die ausgewählte Logikfamilie hängt von der Anwendung ab, für die sie benötigt wird. Einige der Einflüsse, die bei der Auswahl einer Logikfamilie berücksichtigt werden müssen, sind Geschwindigkeit, Kosten und Leistungsaufnahme.

19 Flip-flops

A digital system needs memory elements to store digital information. One type of memory element is the bistable multivibrator commonly called a latch. A latch is an electronic device that has two stable states. It remains indefinitely in one of these states until it is triggered into the other state.

The SR latch

An SR latch can be constructed from two cross-coupled NAND gates, as shown in Fig 19.1. The feedback connections ensure that the latch can exist only in one of two stable states, either $Q = 1$ ($\overline{Q} = 0$) called the 1 state, or $Q = 0$ ($\overline{Q} = 1$) called the 0 state. The latch can be controlled by using the inputs S and R which are termed the set and reset inputs. The output corresponding to various states at the inputs is shown in the table of Fig 19.1.

When S and R are both 1, no change occurs and the circuit remains in the last latched state. When S is 1 and R is 0, Q becomes 1. When S is 0 and R is 1, Q becomes 0. When R and S are both 0, then the state of Q is ambiguous and unpredictable. Such a condition should be avoided.

The clocked SR flip-flop

In digital systems, it is usually necessary that data be entered at a definite time. The timing of the system is done by using a regular sequence of pulses from a clock. A latch that can change state only during a clock pulse is called a flip-flop. The circuit diagram and truth table for a clocked SR flip-flop is shown in Fig 19.2. It will be seen that two controlling gates have been added to the circuit of Fig 19.1. The clock drives both gates.

Between clock pulses (Ck = 0), the outputs of the controlling gates are both 1, and are independent of the states of R and S. Hence there is no change in the output Q. When a clock pulse is present (Ck = 1), the output is as shown in the truth table of Fig 19.2. If Ck = 1, S = 1, and R = 1, then the outputs of the controlling gates are both 1 and the output is unpredictable.

The JK flip-flop

The ambiguity in the truth table of the SR flip-flop (when S = R = 1) is removed in the JK flip-flop the circuit of which is shown in Fig 19.3. The circuit consists of four NAND gates and involves feedback from the outputs Q and $\overline{Q}$ to the input gates. The truth table is also shown in the figure and it will be seen that the first three rows are the same as for the SR flip-flop. The ambiguity in the output for S = R = 1, is replaced by $Q_{n+1} = \overline{Q}_n$ for J = K = 1. If J and K are both in the 1 state, the output is complemented by the clock pulse.

Preset and Clear

When the power is switched on a flip-flop may be in an unknown state, and it is usually necessary to assign a definite state to a flip-flop before starting an operation. This may be accomplished by the addition of preset and clear inputs as shown in the diagram. A clear operation which corresponds to Q = 0 may be performed by making Cr = 1, Pr = 0, Ck = 0.

The feedback applied to the JK flip-flop can make it unstable under certain conditions. This difficulty may be avoided by using a master-slave flip-flop.

The JK master-slave flip-flop

The block diagram of a master-slave flip-flop is shown in Fig 19.4. The diagram shows two flip-flops called the master and the slave, with feedback from the output of the slave being applied to the input of the master. Clock pulses which are applied to the master are inverted before they are applied to the slave. The master is enabled for the time duration of a clock pulse and its operation is in accordance with the JK truth table. The slave is disabled, and cannot change state for the duration of the clock pulse. After the pulse is over, the slave is enabled, and the output of the master is transferred to the output of the slave. Instability and race around conditions are avoided in this circuit.

The D-type flip-flop

The D-type flip-flop is a modified form of the JK flip-flop in which an inverter is included in the input as shown in Fig 19.5. This ensures that K is always the complement of J, and the state corresponding to J = K = 1 cannot arise. The truth table for this is shown in the figure. It will be seen that the output Q_{n+1} after the pulse is equal to the input D_n before the pulse. The D-type flip-flop functions as a delay device because the input at D is transferred to the output at the next pulse.

T-type flip-flop

Yet another kind of flip-flop is shown in Fig 19.6. This is called the toggle or T-type flip-flop, because it behaves like a toggle switch. In this circuit, if J = K = 1, the output changes with each clock pulse as can be seen from the JK truth table.

Vocabulary

accomplish	erfüllen *v*	**hence**	daraus, daher *adv*
ambiguous	zweideutig, unklar *adj*	**indefinitely**	unbegrenzt *adj*
assign	zuordnen *v*	**invert**	umkehren *v*
binary information	Binärinformation *f*	**latch**	selbsthaltender Schalter *m*
bistable multivibrator	bistabile Kippschaltung *f*	**master-slave flip-flop**	bistabiles Master-Slave Kippglied *n*
clear input	Löscheingang *m*	**memory element**	Signalspeicher *m*
cross-couple	querverkoppeln *v*	**preset input**	Vorwahleingang *m*
delay device	Verzögerungsbauelement *m*	**regular**	regelmaßig, normal *adj*
disable	sperren *v*	**sequence**	Folge *f*
drive	treiben *v*	**stable state**	Ruhezustand *m*
enable	in Betrieb setzen, aktivieren *v*	**toggle switch**	Kippschalter *m*
flip-flop	Flip-Flop, bistabiles Kippglied *n*		

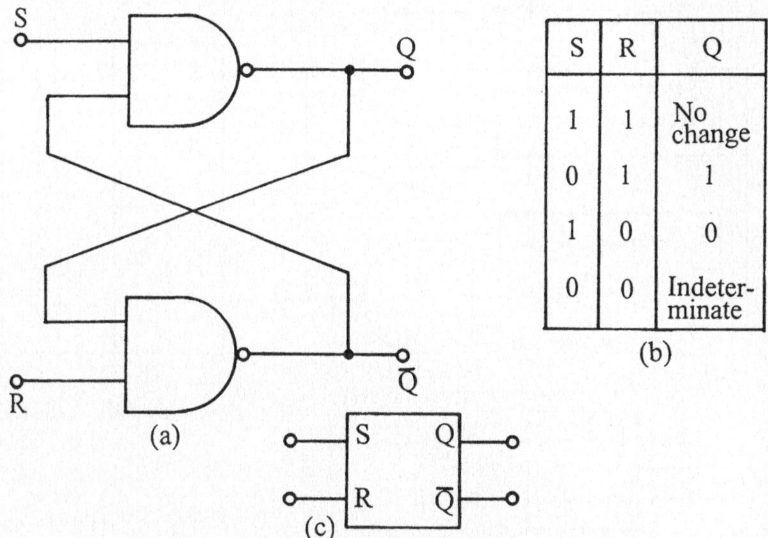

S	R	Q
1	1	No change
0	1	1
1	0	0
0	0	Indeterminate

Fig 19.1 (a) An SR latch constructed of NAND gates, (b) truth table, (c) block diagram

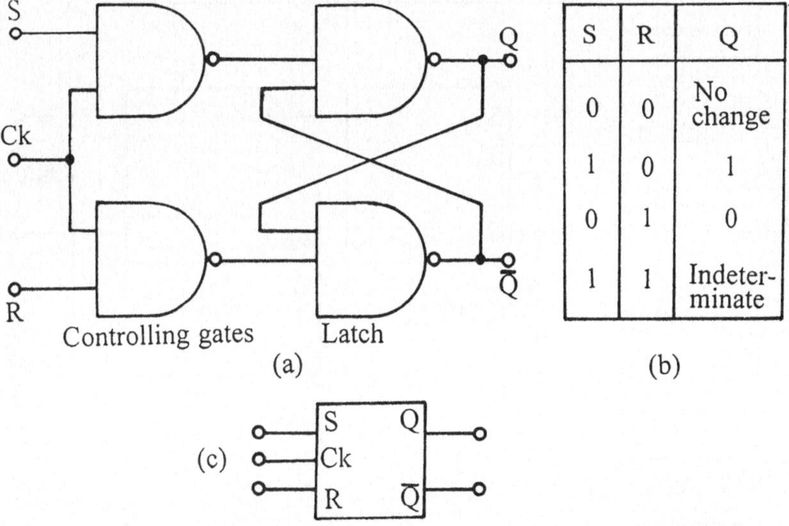

S	R	Q
0	0	No change
1	0	1
0	1	0
1	1	Indeterminate

Fig 19.2 (a) A clocked SR flip flop, (b) truth table, (c) block diagram

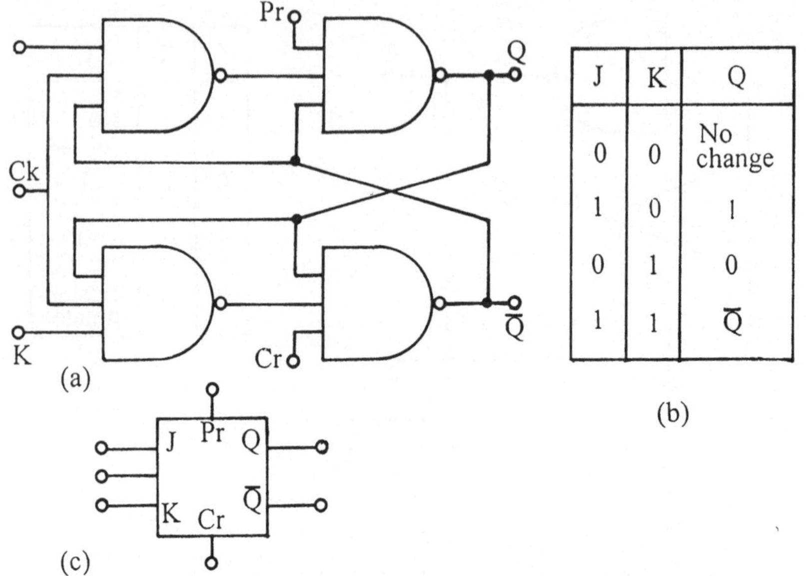

J	K	Q
0	0	No change
1	0	1
0	1	0
1	1	$\overline{Q}$

Fig 19.3 (a) A JK flip-flop, (b) truth table, (c) logic symbol

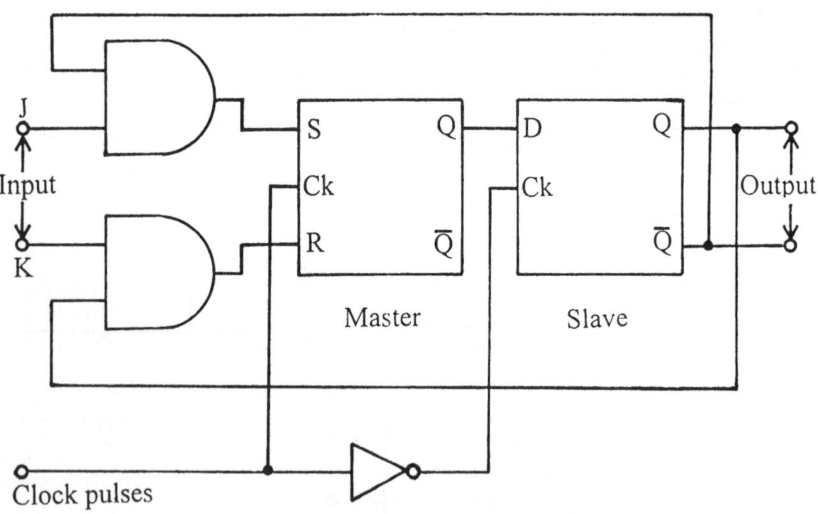

Fig 19.4 A JK master-slave flip-flop

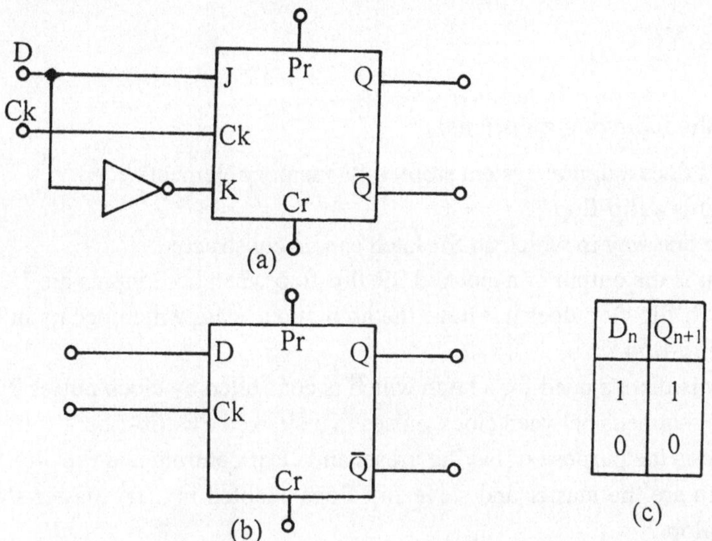

D_n	Q_{n+1}
1	1
0	0

Fig 19.5 (a) A D-type flip-flop constructed from a JK flip-flop,

(b) logic symbol, (c) truth table

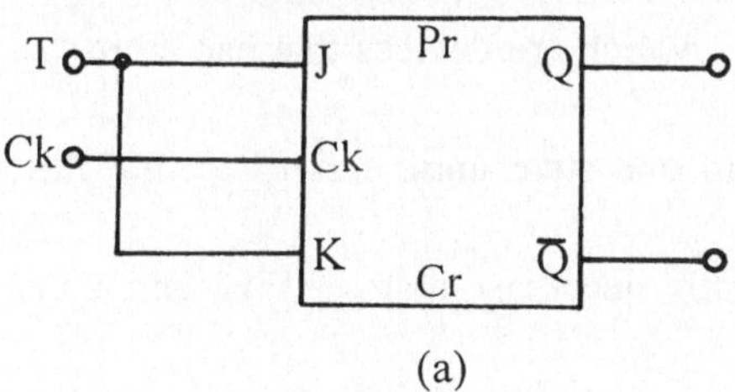

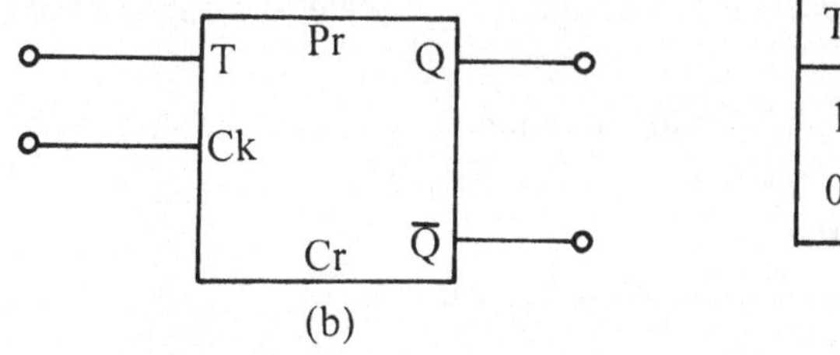

T_n	Q_{n+1}
1	$\overline{Q}_n$
0	Q_n

(c)

Fig 19.6 (a) A T-type flip-flop constructed from a JK flip-flop,

(b) logic symbol, (c) truth table

Exercises XIX

1. Answer the following questions:

a) What does a digital system store in its memory elements ?
b) What is a flip-flop ?
c) Give one way in which an SR latch can be constructed.
d) What is the output of a clocked SR flip-flop when both inputs are 1 ?
e) Which flip-flop does not have the ambiguous state which occurs in the SR flip-flop ?
f) Why is there a need for a latch which is controlled by clock pulses ?
g) What happens between clock pulses in a clocked SR flip-flop ?
h) What is the purpose of having preset and clear controls in a flip-flop ?
i) When are the master and slave flip-flops enabled in a JK master-slave flip-flop ?
j) What are the main characteristics of the T-type flip-flop ?

2. Fill in the gaps in the following sentences:

a) A digital system needs ____ which are devices that can store ____ information.
b) A flip-flop remains ____ in one state until it is ____ into the other state.
c) In digital systems it is usually necessary that ____ be entered in at a ____ time.
d) A ____ that can change state ____ is called a flip-flop.
e) When the ____ is switched on, a flip-flop may be in an ____.
f) It is usually necessary to assign a ____ state to a flip-flop before starting ____.
g) The master is ____ for the duration of ____ and the slave is disabled.
h) Clock pulses which are ____ to the master are ____ before being applied to the slave.
i) The D-type flip-flop is a modified form of the ____ in which an ____ is included in the input.
j) The D-type flip-flop functions as a ____ because the input at D is transferred to the output at the ____.

3. Translate into English

a) Ein digitales System benötigt Speicherbausteine, die Binärinformationen speichern können. Einer der einfachsten Speicherbausteine ist die bistabile Kippschaltung, auch Flip-Flop genannt.

b) Ein Flip-Flop ist ein elektronisches Bauelement, das zwei stabile Zustände besitzt. Es verweilt unbegrenzt in einem dieser Zustände, bis es in den anderen Zustand getriggert wird.

c) Wenn die Spannungsversorgung eingeschaltet wird, kann sich ein Flip-Flop in einem unbestimmten Zustand befinden, aufgrund dessen es üblicherweise nötig ist, dem Flip-Flop einen eindeutigen Zustand zuzuordnen, bevor der Betrieb aufgenommen wird. Dies kann unter Verwendung der Preset und Clear Eingänge des Flip-Flops vorgenommen werden.

d) Clock-Signale, die an das Master Flip-Flop angelegt werden, werden invertiert, bevor sie an das Slave Flip-Flop angelegt werden. Das Master Flip-Flop ist für die Zeitdauer des Clock-Pulses freigegeben, während das Slave Flip-Flop gesperrt ist und während der Dauer des Clock-Pulses seinen Zustand nicht ändern kann.

e) Das Flip-Flop vom D-Typ ist eine modifizierte Form des JK Flip-Flops, in dem ein eingangsseitiger Inverter eingefügt ist, wie in der entsprechenden Abbildung dargestellt. Dieser stellt sicher, daß K stets das Komplement zu J bildet und ein zweideutiger Ausgangszustand vermieden wird.

20 Shift registers

A flip-flop can store a single binary digit. The storage of a number of binary digits can be accomplished by combining a number of flip-flops to form a device called a shift register. The name shift register originates from the fact that this device takes in one new digit for each clock pulse while at the same time shifting the existing digits by one stage to make room for the new digit.

There are many ways in which data can be entered into and extracted from a shift register. Among the possibilities are, Serial-input Serial-output, Parallel-input Parallel-output, Serial-input Parallel-output, and Parallel-input Serial-output.

Serial-input Parallel-output

A four bit shift register is shown in Fig 20.1. The flip-flops are first cleared using the Cr and Pr inputs so that all the Q outputs are zero. Suppose that the digits 1001 have to be stored in the shift register. The input state will initially be at 1, corresponding to the least significant bit (LSB) which is 1. The output at Q_1 will become 1 after the first clock pulse. The input will now change to 0. At the next clock pulse, the 1 in Q_1 will be transferred to Q_2, while the 0 at the input of FF1 will now be at the output Q_1. In this way the input bits will be entered into the register progressively, and after four clock pulses the four data bits will be on the four outputs of the flip-flops. Since each digit is now available on a separate Q output, the bits can be read simultaneously to obtain a parallel output.

Serial-input Serial-output

The entering of the data is done serially in the same way as described above. The data can then be removed serially from the output terminal Q_4. The LSB is first removed, and then the other bits are removed one at a time after each successive clock pulse.

Parallel-input Serial-output

The bits can be loaded into the shift register simultaneously using the preset (Pr) inputs. Once this has been done, they can be read out serially from the terminal Q_4 in the same way as described above.

Parallel-input Parallel-output

The bits are entered simultaneously using the preset (Pr) inputs, and then removed simultaneously at a selected time from the Q_1 to Q_4 outputs.

Right-Left shift register

Shift registers can be fitted with gates which allow the data bits to be shifted to the left or to the right.

Multiplication and division

A left-right shift register can be used to perform multiplication and division by two. If a clock pulse is applied to a right shift register, each bit is shifted to the next lowest significant place and this shift is equivalent to a division by two. If a clock pulse is applied to a left shift register, each bit is moved to the next higher significant place which is equivalent to a multiplication by two.

Digital delay line

A shift register may be used as a time delay device. An input train of data pulses which enters an n-stage shift register appears as a pulse train at the output after a time (n - 1)T, where T is the clock period.

Ring counter

If the last Q output terminal of the shift register is connected to the first input terminal, then data (which can be entered in parallel form) will circulate round the register when clock pulses are applied and not flow out of the register. Such an arrangement is called a ring counter.

One application for a ring counter is its use as a replacement for the distributor in an automobile engine. For example a four cylinder engine would need a ring counter with four flip-flops. Instead of a mechanical cam opening and closing the contacts to produce an ignition spark, clock pulses are produced by a sensor on the engine flywheel. These pulses are used to shift a logic 1 round the four stage ring counter. The phase of the pulses can be varied so that the logic 1 arrives at exactly the right time to fire the engine.

Vocabulary

adjust	einstellen, regulieren *v*	**initially**	zuerst, anfänglich adv
arrangement	Anordnung *f*	**least significant bit (LSB)**	geringstwertiges Bit n
binary digit	Binärzeichen, Bit *n*	**most significant bit (MSB)**	höchstwertiges Bit *n*
circulate	kreisen, umlaufen *v*	**progressively**	fortlaufend *adj*
clock pulse	Taktimpuls *m*	**sensor**	Meßfühler m, Sensor *m*
correspond to	beziehen auf, passen zu *v*	**separate**	einzeln, getrennt *adj*
distributor	Verteiler *m*	**sequentially**	nacheinander *adv*
exactly	genau *adv*	**serial**	reihenweise *adj*
exist	existieren, vorhanden sein *v*	**shift register**	Schieberegister *n*
extract	herausziehen, extrahieren *v*	**simultaneously**	gleichzeitig *adv*
flywheel	Schwunggrad *n*	**store**	speichern, lagern *v*
ignition spark	Zündfunke *m*	**successive**	aufeinanderfolgend *adj*
		train of pulses	Folge von Impulsen *f*

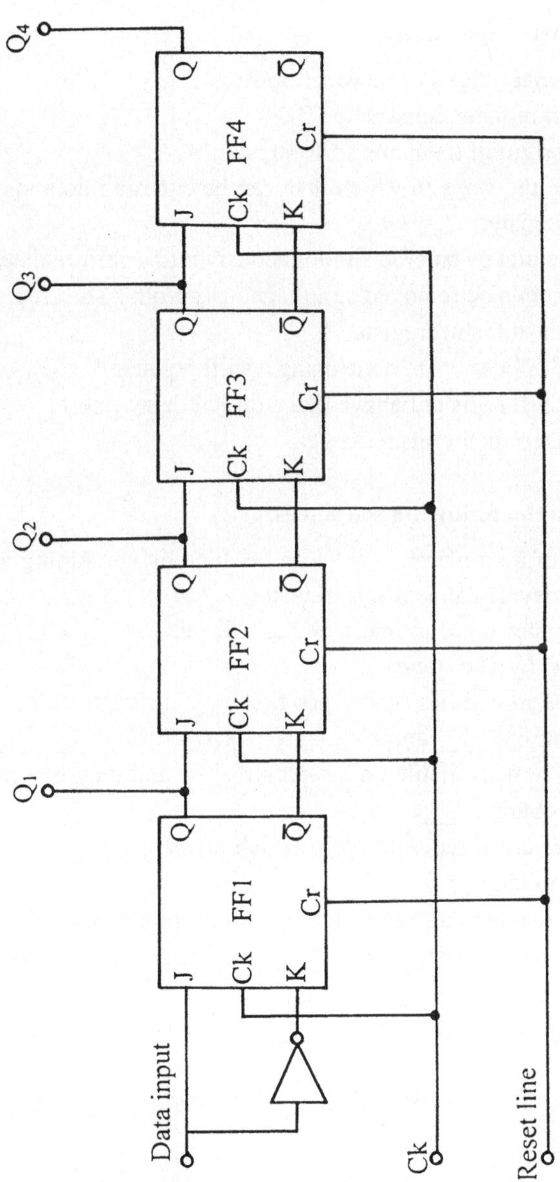

Fig 20.1 A four bit shift register

Exercises XX

1. Answer the following questions:

a) How many binary digits can a flip-flop store ?
b) How is a shift register constructed ?
c) What is the origin of the name shift register ?
d) State some of the ways in which data can be entered into and extracted from a shift register.
e) How can data bits be entered simultaneously into a shift register ?
f) How can data bits be removed simultaneously from a shift register ?
g) What is a right-left shift register ?
h) How can multiplication be done using a shift register ?
i) Why does a shift register behave like a digital delay line ?
j) How is a ring counter constructed ?

2. Fill in the gaps in the following sentences:

a) The storage of a number of ____ can be done by combining a number of ____ in a device called a shift register.
b) The shift register takes in one new digit for each ____ while ____ the existing digits by one stage.
c) The flip-flops in a shift register are first ____ using the Cr and Pr inputs so that all the ____ are 0.
d) Each digit is now available on a separate ____ and may be read simultaneously to obtain a ____.
e) Shift registers are fitted with ____ which allow the ____ to be shifted to the left or to the right.
f) A ____ shift register can be used to ____ multiplication or division.
g) If a clock pulse is applied to a ____, each bit is moved to the next higher ____.
h) An input train of pulses entering an ____ appears as a pulse train ____ (n - 1)T.
i) A ring counter may be constructed by connecting ____ of a shift register to the ____.
j) One ____ for a ring counter is its use as a ____ for the distributor in an automobile engine.

3. Translate into English:

a) Ein Flip-Flop vermag eine einzelne Binärstelle zu speichern. Die Speicherung mehrerer Binärstellen kann durch Verbinden mehrerer Flip-Flops zu einem Schieberegister genannten Baustein erreicht werden.

b) Der Name Schieberegister rührt daher, daß dieser Baustein eine neue Binärstelle pro Taktpuls hereinnimmt, wobei die existierenden Binärstellen um eine Stufe verschoben werden, um für die neue Stelle Platz zu schaffen.

c) Schieberegister können mit Gattern versehen werden, die es ermöglichen, die Datenbits nach links oder rechts zu verschieben. Ein links-rechts Schieberegister kann zur Ausführung von Multiplikationen oder Divisionen durch zwei eingesetzt werden.

d) Ein Schieberegister kann als Zeitverzögerungsbaustein benutzt werden. Wird eine Folge von Eingangsdatenimpulsen in ein n-stufiges Schieberegister eingelesen, so erscheint diese nach einer Zeit (n - 1)T als Folge von Impulsen am Ausgang, wobei T die Taktfrequenz ist.

e) Eine Anwendung eines Ringzählers ist seine Verwendung als Platzhalter für den Verteiler in einem Automobilmotor. Ein Auto mit einem Vierzylindermotor würde einen Ringzähler mit vier Flip-Flops benötigen.

21 Electronic counters

Electronic counters are widely used in instruments, computers, and other digital systems. The availability of counters in IC form has made electronic counting a reliable and inexpensive process. Electronic counters usually use the binary system, but the binary output can easily be converted into decimal or other form, by the use of suitable converting circuits. Both binary and decimal counters are available as IC chips. The basic element of the binary counter is the master-slave flip-flop set to toggle or reverse its state on each clock pulse.

Asynchronous or ripple counters

A ripple counter consists of a series of flip-flops the output Q of each flip-flop being connected to the clock input of the next one as shown in Fig 21.1. The J and K input terminals of all the flip-flops are connected to the supply voltage so that J = K = 1. In this way each stage is converted into a T-type flip-flop. All outputs are initially set to zero using the common reset line.

The pulses which are to be counted are applied to the clock input of the first flip-flop. In a toggle type flip-flop, the master changes state when the pulse at its clock input changes from 0 to 1. The state of the master is transferred to the slave when the pulse falls from 1 to 0. Therefore it follows that:

1. the output of the first flip-flop changes state at the falling edge of the input pulse.

2. the outputs of the other flip-flops change state when the output of the preceding flip-flop changes from 1 to 0.

This means that the first flip-flop divides the number of pulses at the input by 2, the second by 4, and the nth flip-flop by 2^n. A counter with n flip-flops will revert to its original state after a count of 2^n. The number of pulses that have arrived at the input can be obtained in binary form by reading the outputs of all the flip-flops. Here the first flip-flop corresponds to the LSB and the last flip-flop to the MSB.

The states of the flip-flops corresponding to the number of input pulses is shown in the table of Fig 21.2. The waveforms of the four outputs of a four stage ripple counter are shown in Fig 21.3. This type of counter is called a ripple counter, because each stage responds only after the previous stage has completed its transition. Thus the clock pulse ripples through the chain.

The counter delay time involved in the counting process, is effectively equal to the sum of the propagation delay times of all the flip-flops. The counter becomes ineffective when the counter delay time is longer than the interval between pulses. Higher counting rates are possible by using synchronous counters.

Decade counters

It is very often necessary to count to a base N, which is not a multiple of two. For example in the decimal system it is necessary to count to the base ten. The diagram in Fig 21.4 shows how a four-stage binary ripple counter can be modified to become a decade counter. Such a counter would normally revert to its original state after a count of sixteen. The addition of a gate as shown in the diagram ensures that the counter reverts to its original state after a count of ten as explained below.

The decimal number 10 corresponds to the binary number 1010 (LSB). At the count of 10, the outputs of the flip-flops will be $Q_0 = 0$, $Q_1 = 1$, $Q_2 = 0$, $Q_3 = 1$. This means that only Q_1 and Q_3 need to be reset to start the counting sequence. The outputs Q_1 and Q_3 are taken to a NAND gate whose output feeds all the clear inputs in parallel. After the tenth input pulse, the output of the NAND gate becomes 0 and all the flip-flops are reset to 0.

Synchronous counters

In a ripple counter, the time required for it to respond to an input pulse, is approximately equal to the sum of the propagation delay times of all the flip-flops. In a synchronous counter all the flip-flops are clocked synchronously (simultaneously) by the input pulses, with the result that the propagation delay times are considerably reduced. The maximum rate of counting of a synchronous counter can be twice as high as that of an asynchronous counter.

The operation of a synchronous counter is based on the principle that each stage toggles only on clock pulses that occur when the outputs of all less significant stages are 1. The block diagram of a four stage divide by 16 synchronous counter is shown in Fig 21.5. The counter operates as follows:

1. The first flip-flop toggles on each clock pulse because J = K = 1.
2. In accordance with the synchronous principle mentioned above, the second flip-flop toggles only when the first stage output is 1. This happens because Q_1 is connected to the J and K inputs of stage 2.
3. The third flip-flop should toggle only when Q_1 and Q_2 are both 1. This is ensured by connecting Q_1 and Q_2 to an AND gate whose output is connected to J_3 and K_3.
4. The fourth flip-flop should toggle when Q_1, Q_2 and Q_3 are each 1. This is accomplished by using a second AND gate to produce a 1 output at J_4 and K_4.

Vocabulary

asynchronous	asynchron *adj*
based on	gegründet auf
considerably	beträchtlich *adv*
decade counter	Dekadenzähler *m*
ensure	sicherstellen *v*
in accordance with	in Übereinstimmung mit
ineffective	wirkungslos, unwirksam *adj*
inexpensive	nicht teuer, preiswert *adj*
preceding	vorhergehend *adj*
reliable	zuverlässig *adj*
reset line	Rücksetzleitung *f*
respond	reagieren *v*
revert	zurückkehren *v*
ripple counter	asynchroner Zähler *m*
stage	Stufe *f*
supply voltage	Betriebsspannung *f*
transition	Übergang *m*

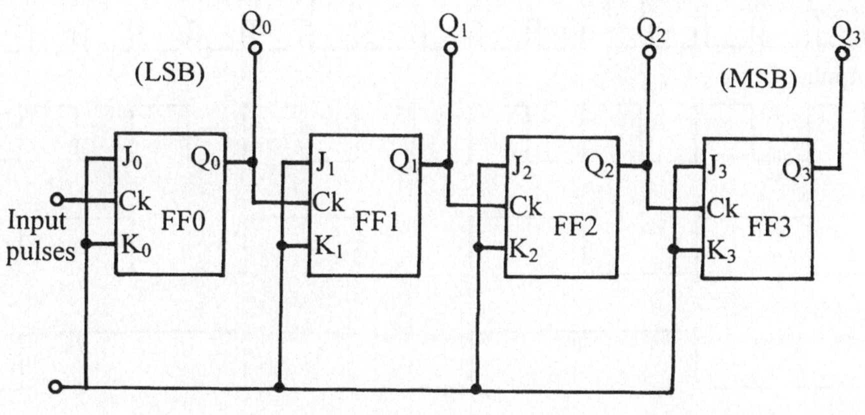

Fig 21.1 A four stage asynchronous or ripple counter

Number of input pulses	FLIP-FLOP outputs			
	Q_3	Q_2	Q_1	Q_0
0	0	0	0	0
1	0	0	0	1
2	0	0	1	0
3	0	0	1	1
4	0	1	0	0
5	0	1	0	1
6	0	1	1	0
7	0	1	1	1
8	1	0	0	0
9	1	0	0	1
10	1	0	1	0
11	1	0	1	1
12	1	1	0	0
13	1	1	0	1
14	1	1	1	0
15	1	1	1	1
16	0	0	0	0

Fig 21.2 Table showing the states of the flip-flop outputs

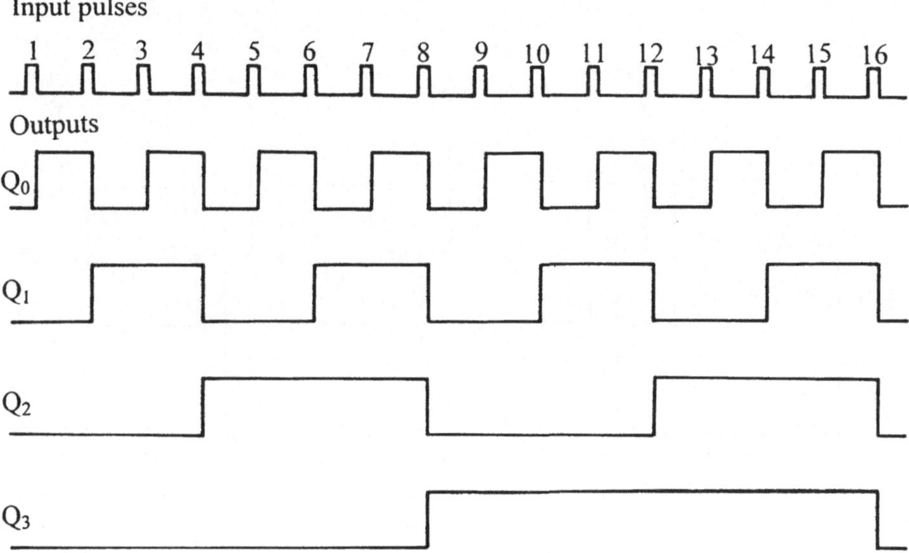

Fig 21.3 Input pulses and output waveforms for a four stage ripple counter

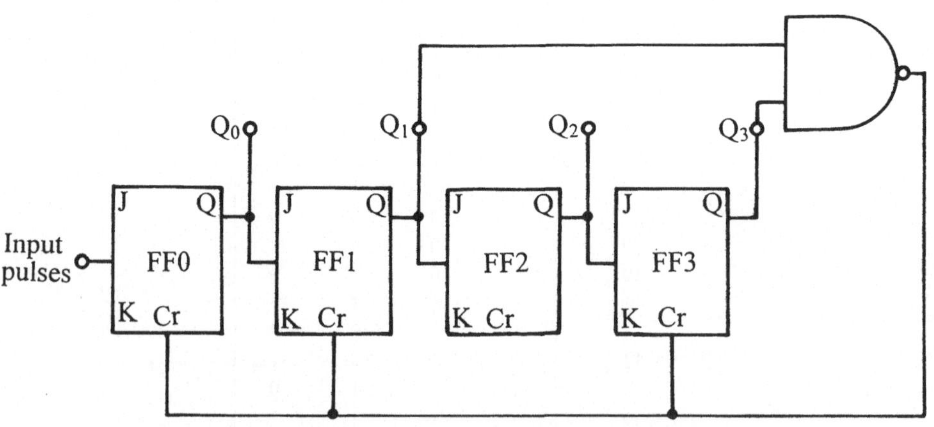

Fig 21.4 A four stage decade counter

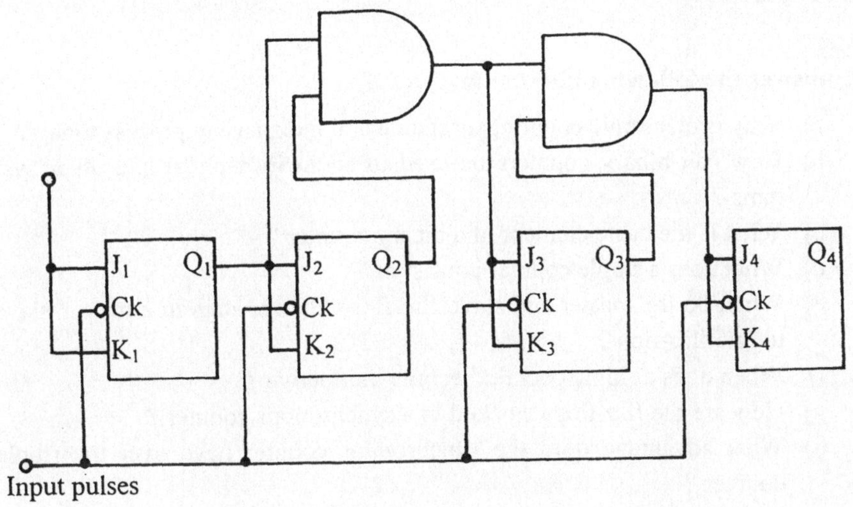

Fig 21.5 A four stage synchronous counter

Exercises XXI

1. Answer the following questions:

a) Why is electronic counting a reliable and inexpensive process today ?
b) How can binary counters be used to count in decimal and other systems ?
c) What is the basic element of a binary counter ?
d) What does a ripple counter consist of ?
e) When do the master and slave flip-flops change state in a master-slave toggle flip-flop ?
f) When does a ripple counter become ineffective ?
g) How are the flip-flops clocked in a synchronous counter ?
h) What advantage does the synchronous counter have over the ripple counter ?
i) Under what conditions does each stage of a synchronous counter toggle ?
j) Why is an asynchronous counter called a ripple counter ?

2. Fill in the gaps in the following sentences:

a) The availability of counters in ____ has made electronic counting a ____ and inexpensive process.
b) The binary output of a counter can be converted into ____ form by the use of ____ circuits.
c) The basic element of a binary counter is the ____ set to ____ on each clock pulse.
d) A ripple counter consists of a series of flip-flops, the ____ of each flip-flop being connected to the ____ of the next one.
e) In a toggle type flip-flop, the ____ changes state when the pulse at its ____ changes from 0 to 1.
f) The first flip-flop divides the number of pulses by ____, and the second by ____.
g) This type of counter is called a ____ because each stage responds only after the ____ has completed its transition.
h) The counter becomes ____ when the counter time delay is greater than the ____.

i) In a synchronous counter, all the flip-flops are _____ synchronously by the ____ .
j) In a synchronous counter, each state ____ only on clock pulses that occur when the outputs of all ____ are 1.

3. Translate into English:

a) Zähler folgen im Normalfalle dem Binärsystem, aber die Binärausgänge können leicht in Dezimalform oder eine andere Form durch Einsatz geeigneter Konverterschaltungen umgesetzt werden.
b) Ein Asynchronzähler besteht aus einer Folge von Flip-Flops, wobei der Ausgang Q jedes Flip-Flops mit dem Clock-Eingang des nächsten verbunden wird. Die J und K Eingänge aller Flip-Flops werden auf Betriebsspannung gelegt, so daß $J = K = 1$ ist.
c) Das erste Flip-Flop dividiert die Anzahl der Pulse am Eingang durch 2, das zweite durch 4 und das n-te durch 2^n. Ein Zähler mit n Flip-Flops kehrt nach dem Zählen von 2^n wieder in den Ausgangszustand zurück.
d) Die Pulse, die gezählt werden sollen, werden an den Clock-Eingang des ersten Flip-Flops angelegt. Alle Ausgänge werden zunächst durch Betätigung des allgemeinen Resets auf 0 gesetzt.
e) Bei einem Asynchronzähler entspricht die für eine Reaktion benötigte Zeit ungefähr der Durchlaufverzögerung aller Flip-Flops. In einem Synchronzähler werden alle Flip-Flops simultan getaktet und somit die Durchlaufverzögerung beträchtlich verringert.

22 Semiconductor memories

In almost all computer and data processing systems it is necessary to be able to store information and retrieve it when required. The storage unit or memory is one of the most active parts of a computer. It stores the program, the input data, and also the processed data, at various stages of the computing process. A computer memory has many thousands of registers, each register storing a word, where a word consists of a number of bits. Virtually all of the memories in use today are of the semiconductor type.

Read only memory (ROM)

A read only memory has a group of registers or memory locations in each of which a word is stored permanently or semipermanently. By applying suitable control signals it is possible to read the word stored in any memory location. The term read refers to the process of making the data stored in the memory to appear at the output terminals.

One of the ways in which a ROM can be built by using diodes is shown in Fig 22.1. Each horizontal row corresponds to a register or memory location. Each register has a different number of diodes in different positions. To read the contents of a memory location, say ML5, the switch on the left is moved to position 5. Since +5V is applied at position 5, the two diodes conduct and the points P_1 and P_2 are raised to level 1. P_0 and P_3 remain at level 0. This results in the output becoming 0110. This is the word stored in the memory location.

Each memory location has 4 bits, and any 4-bit word can be stored in each memory location by adding or removing diodes. In IC ROMs, the diodes are set in place, and the words are stored at the time of manufacture. In both types of ROM discrete and IC, the words are stored permanently.

Memory addresses

Each word stored in the memory has two parameters, its address or memory location, and the data which is contained in the word. The simple ROM discussed above has eight memory locations and can be addressed by turning a switch. Before the contents of a memory location can be read, it is necessary to excite the particular address line (elevate to level 1) corresponding to the particular memory location selected.

In practice the selection of memory locations is done electronically by the use of a binary addressing system. One way of doing this for the ROM previously considered is shown in Fig 22.3. The address of the memory location to be selected is applied to the terminals $A_0A_1A_2$. Each address word causes a high voltage to be applied to only one of the lines corresponding to the selected memory location. For example if the address word is 110, the contents of the memory location 6 will be read. Such a device is called a decoder and functions like a multiposition switch.

Programmable ROM (PROM) and erasable programmable ROM (EPROM)

In a ROM, the data stored in each memory location is fixed at the time of manufacture. Mass produced ROMs are usually manufactured in integrated form by the IC manufacturer, and not by the user.

A programmable ROM (PROM) allows the user to store the data by using an instrument called a PROM programmer. The programmer burns open fusible links at certain bit locations by passing high currents. By this method, the program and data are said to be burnt in and stored permanently (nonerasable).

The erasable PROM uses MOSFETs to store data with the help of a programmer. The stored data may be removed or erased when necessary by the use of ultraviolet light.

Random access memory (RAM)

In many types of digital systems, it is necessary to store data temporarily, and retrieve it at a later time. In a random access memory, digital data can be stored or removed from any chosen location. Since data can be written into or read out of each memory location at random such a system is called a read/write memory in contrast to a read only memory or ROM. Random access memories are almost invariably made in the form of IC chips. Semiconductor random access memories are volatile, meaning that all stored information is lost when a power failure takes place.

The low power consumption of CMOS memory cells makes the use of stand by battery power economical in certain applications, with the result that stored information can be preserved even in case of a power failure.

Static RAM

Static RAMs use bipolar or MOS latches as storage devices. An MOS latch used in a static RAM is shown in Fig 22.2. The circuit action is similar to the latches discussed before with T_3 and T_4 acting as active loads. Such a latch remains indefinitely in a given state, as long as power is supplied to it.

Dynamic RAM

A static RAM has a relatively large number of transistors per unit cell. In a dynamic RAM, this number can be reduced and more memory cells can be packed into in an IC chip of a given size. One type of memory cell used in a dynamic RAM is shown in Fig 22.4. The storage element is a capacitor, and a single transistor is used as a transmission gate to charge the capacitor. The disadvantage of the dynamic RAM cell is that the capacitor loses its charge due to leakage. Additional circuitry is required to refresh the stored information periodically.

Single bit read/write memory

The operation of a single bit read/write memory can be understood by considering the circuit shown in Fig 22.5 The circuit uses a flip-flop as a storage element, and has data input and data output lines.

To read data out of the flip-flops, the address lines must be excited (X = 1). To write into the flip-flops, both the address line and the write enable line must be excited. Then if the write input is made 1, S and Q both become 1. If the write input is made 0, S and Q both become 0.

RAM organization

Consider the organization of a 1 kilobyte RAM. Such a memory is organized in the form of 1024 words or bytes of eight bits each. Each package of 1024 memory elements is organized in a 32 x 32 element array, each element storing one bit of one word. Eight packages of this type are required, one for each bit of the eight bits in a word.

Two dimensional addressing is used as shown in Fig 22.6, and each memory element is identified by the X-Y number of a rectangular matrix. To read or write into a memory element say 2-3, an X decoder identifies row 2 and a Y decoder identifies column 3. Since there are 32 rows and 32 columns, each decoder has 5 ($32 = 2^5$) address inputs, and 32 outputs. Eight such arrays are required as stated before, and the decoders will feed eight arrays in parallel.

Vocabulary

address Addresse *f*
build aufbauen *v*
contain enthalten *v*
content Inhalt *m*
contrast Gegensatz, Kontrast *m*
control system Steuer-, Regelsystem *n*
decoder Dekodierer *m*
discuss besprechen, diskutieren *v*
dynamic RAM dynamisches RAM
EPROM löschbarer PROM *m*
excite anregen, erregen *v*
group Gruppe *f*
identify erkennen *v*
inactive untätig *adj*
location Stelle *f*, Lage *f*
memory Speicher *m*
necessary notwendig *adj*
package (Ver-) Packung *f*
preserve bewahren *v*
PROM programmierbarer Festwertspeicher *m*
RAM Direktzugriffspeicher *m*
Read/Write memory Lese-/Schreibspeicher *m*
require erfordern *v*
retrieve wiederherstellen *v*
transmission gate Steuergatter, Durchlaßgatter *n*
various mehrere, verschiedene *adj*
volatile energieabhängig, flüchtig *adj*

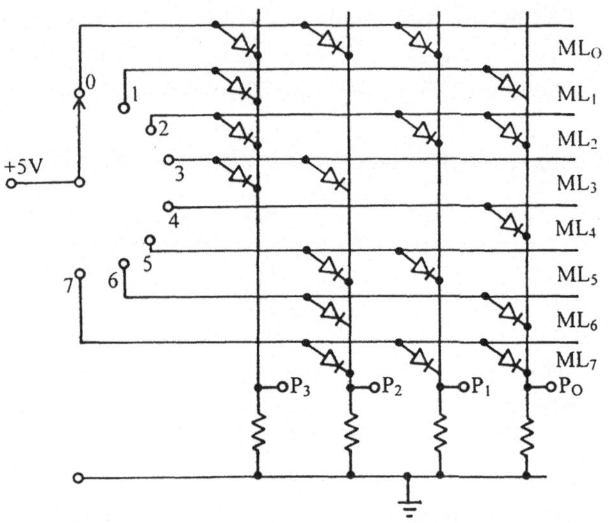

Fig 22.1 A simple diode ROM

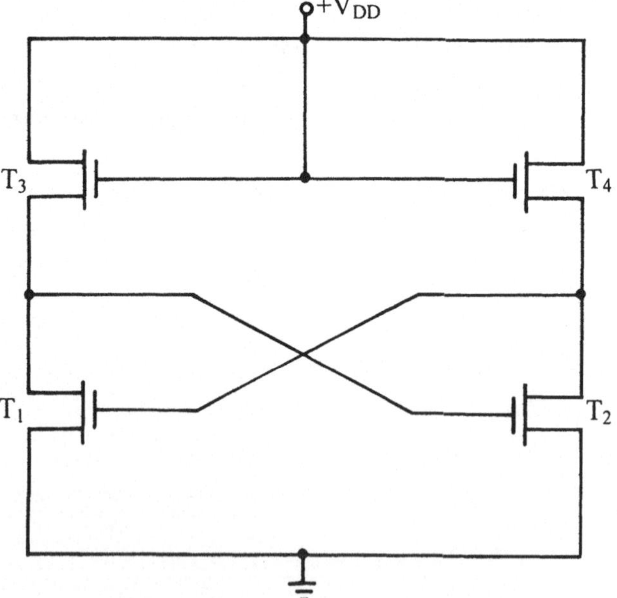

Fig 22.2 MOS latch using active loads

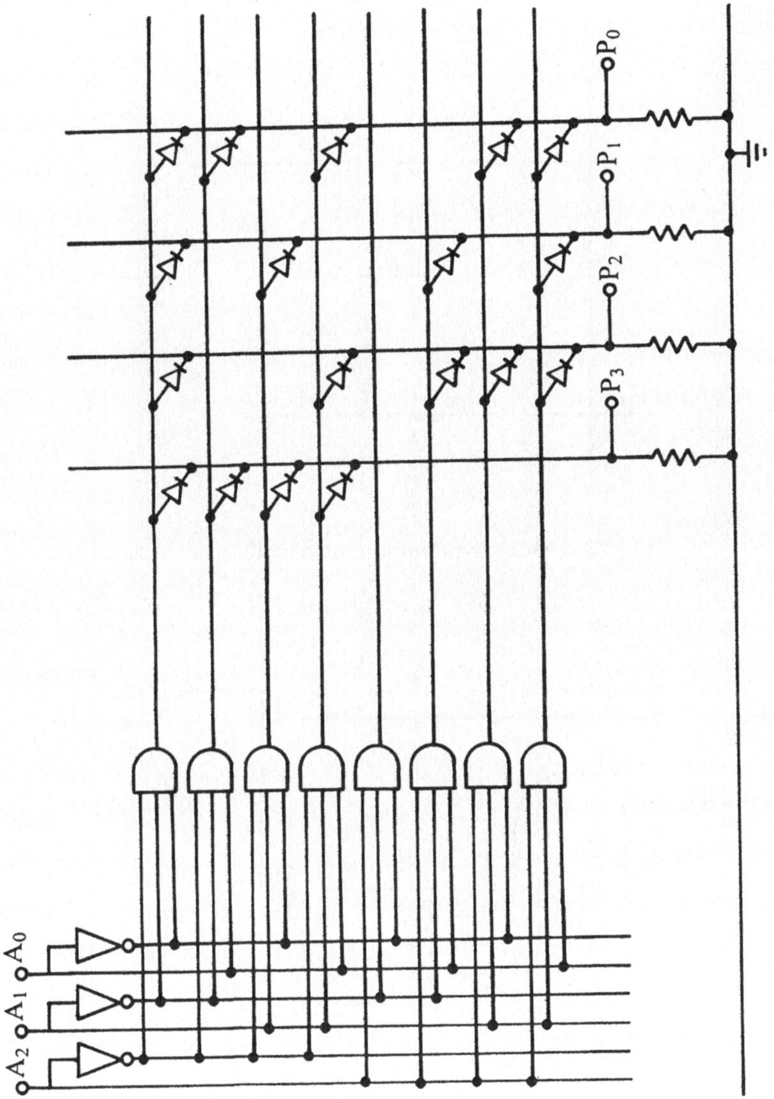

Fig 22.3 ROM IC with built-in decoder

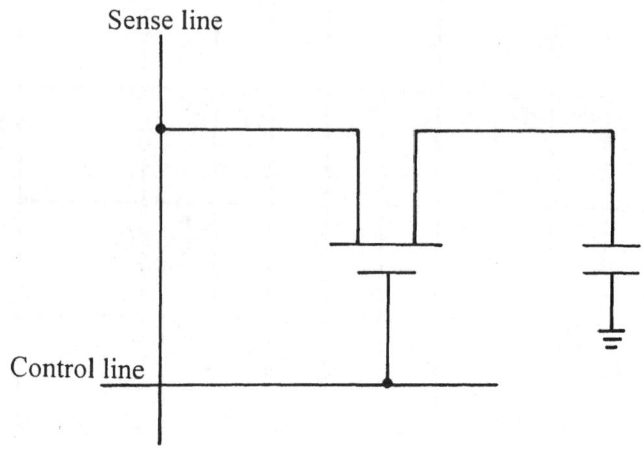

Fig 22.4 MOS dynamic memory cell

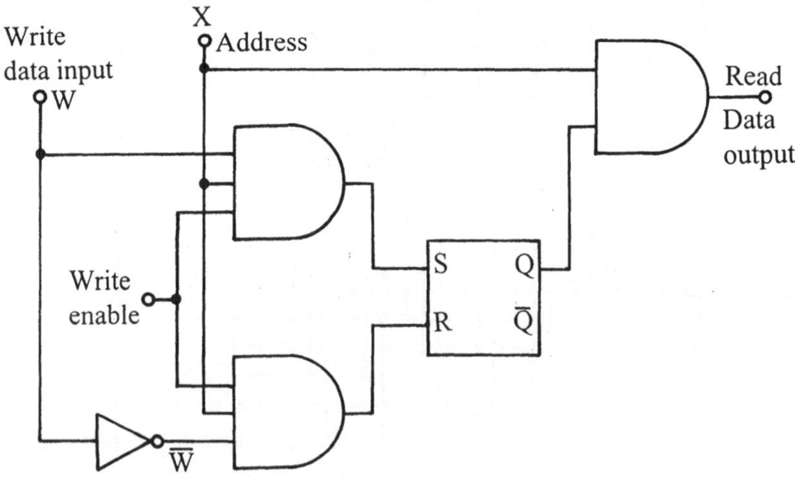

Fig 22.5 A single bit read/write memory

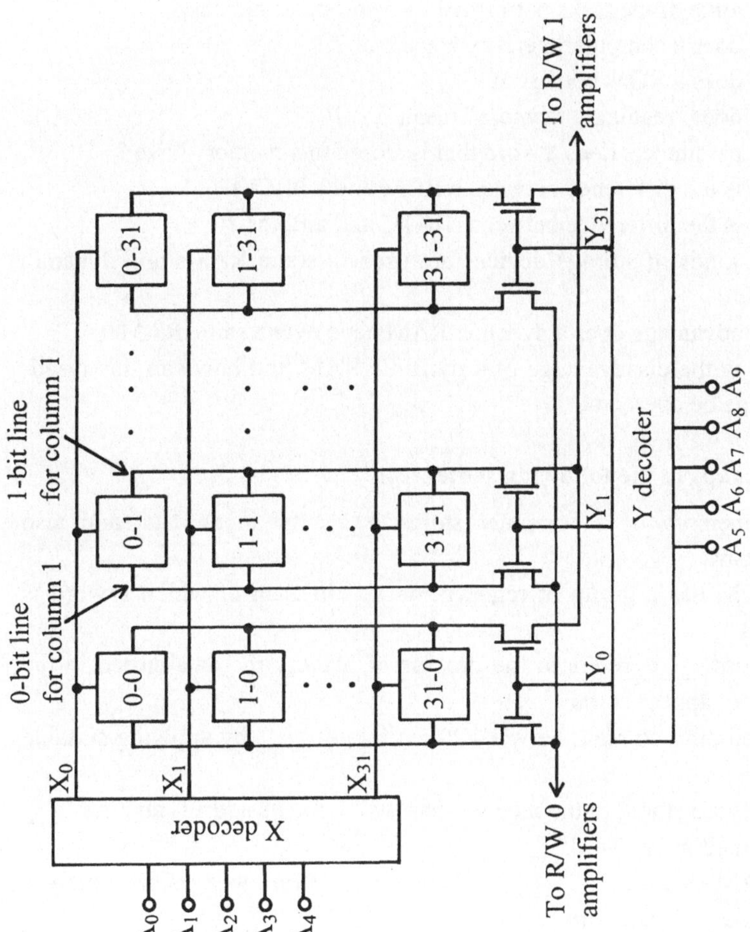

Fig 22.6 Two-dimensional adressing scheme for a RAM

Exercises XXII

1. Answer the following questions:

a) State some of the tasks performed by a computer memory.
b) What does a computer memory consist of ?
c) What does a ROM consist of ?
d) What does "reading a memory" mean ?
e) What parameters does a word that is stored in a memory have ?
f) What is the difference between a ROM and a PROM ?
g) What is the difference between a ROM and a RAM ?
h) What kinds of storage devices are used in static RAMs and dynamic RAMs ?
i) What advantage does a dynamic RAM have over a static RAM ?
j) What is the disadvantage of a dynamic RAM and how can this disadvantage be overcome ?

2. Fill in the gaps in the following sentences:

a) The memory of a computer stores ____, the input data, and also ____data.
b) A ROM has a group of registers or ____in each of which a word is stored ____.
c) The term ____ refers to the process of making the data stored in the memory appear at the ____.
d) It is possible to read the word stored in any ____ by applying suitable ____.
e) Each word stored in the memory has two ____, its address and the ____ contained in the word.
f) A RAM is called a _____ memory, in contrast to a ROM which is called a ____ memory.
g) A PROM allows the ____ to store the data by using an instrument called a ____.
h) In a RAM, digital data can be ____ or removed from any chosen ____.
i) Semiconductor RAMs are ____, meaning that all stored information is lost when a ___ takes place.

j) The low ____ of CMOS memory cells makes the use of ____ economical in certain applications.

3. Translate into English:

a) In Computer-, Steuer-, und Informationssystemen ist es nötig, Informationen jederzeit speichern und wiederherstellen zu können. Der Speicher ist einer der aktivsten Teile eines Computers, der das Programm, die Eingangsdaten und die erzeugten Daten zu verschiedenen Zeitpunkten des Verarbeitungsprozesses speichert.
b) Ein ROM besitzt eine Gruppe von Registern oder Speicherplätzen, in denen jeweils ein Wort fest gespeichert wird. Durch das Anlegen geeigneter Steuersignale ist es möglich, das an einem beliebigen Speicherplatz gespeicherte Wort auszulesen.
c) Die auf jedem Speicherplatz in einem ROM gespeicherten Daten werden zum Zeitpunkt der Herstellung festgelegt. Massenproduzierte ROMs werden normalerweise in Form integrierter Schaltungen vom IC-Hersteller und nicht vom Anwender hergestellt.
d) Direktzugriffsspeicher werden nahezu immer in Form integrierter Schaltungen hergestellt. Halbleiter Direktzugriffsspeicher sind energieabhängig, was heißt, daß alle gespeicherten Informationen verlorengehen, falls ein Ausfall der Spannungsversorgung stattfindet.
e) Das Speicherelement in einem dynamischen RAM ist ein Kondensator, und ein Transistor wird als Durchlaßgatter eingesetzt, um den Kondensator aufzuladen. Der Nachteil der dynamischen RAM-Zelle besteht darin, daß der Kondensator aufgrund von Leckströmen seine Ladung verliert.

Answers to exercises

Exercises I

1.

a) Electrical materials which are poor conductors of electricity are called insulators.
b) Semiconductor materials are important because they play a very useful role in the field of electronics.
c) Allowed bands refer to the restricted ranges of energy that electrons in a solid can have.The allowed bands are separated by ranges of energy which electrons cannot have, called forbidden bands.
d) A completely filled band does not make any contribution to the conductivity of a solid.
e) A partially filled conduction band makes a solid a good conductor.
f) An insulator has energy bands which are either completely filled or completely empty at room temperature.
g) The highest filled band is called the valence band.
h) An insulator and a semiconductor have the same type of band structure at absolute zero.
i) The distinguishing properties of a semiconductor are its small conductivity and the fact that its conductivity increases with temperature.
j) Elements which have one valence electron have a conduction band which is only half-filled with electrons and are therefore good conductors of electricity.

2.

a) The electrical conductivity of a solid depends on its band structure.
b) The energies of the electrons in a solid can only lie within certain restricted ranges called allowed bands.
c) Electrons in a completely filled band do not make a contribution to the conductivity of a solid.

d) A solid in which some bands are completely filled, and others completely empty is an insulator.
e) The highest filled band is called the valence band, and the next higher band is called the conduction band.
f) The bandgap between the valence and conduction bands is large for an insulator.
g) At absolute zero, the valence band of a semiconductor is completely filled, and the conduction band is completely empty.
h) Elements which have only one valence electrons are metals, because the conduction band is only half full.
i) Elements which have two valence electrons can also behave like metals, when their valence and conduction bands overlap.
j) The increase in conductivity with temperature is one of the distinguishing features of a semiconductor.

3.

a) Materials having a high electrical conductivity are termed good conductors. Materials which have poor conducting properties are called insulators. The conductivity of a semiconductor lies between that of a conductor and that of an insulator.
b) The electrical conductivity of a solid depends on its energy band structure.The energy of the electrons can only lie within certain allowed ranges called allowed bands.The allowed bands are separated by ranges of energy which the electrons cannot have, called forbidden bands.
c) The highest filled band is called the valence band and the next higher band is called the conduction band. The bandgap between the valence and conduction bands is of the order of 5 eV for an insulator.
d) At a temperature of 0 K, a semiconductor has a band structure that is identical to that of an insulator, and it behaves like an insulator. At room temperature electrons are transferred from the valence band into the conduction band.
e) When the temperature is increased the conductivity of a semiconductor is increased, which is the opposite of the behaviour of a metal. This increase in conductivity with temperature is one of the distinguishing properties of a semiconductor.

Exercises II

1.

a) Pure semiconductors are called intrinsic semiconductors and impure semiconductors are called extrinsic semiconductors.
b) Elements like silicon and germanium are semiconductors. Among the compounds that are semiconductors are some intermetallic compounds and some metallic oxides.
c) The atomic bonds in a silicon crystal are covalent bonds. This type of bond results in a stable configuration in which each atom is surrounded by eight electrons.
d) When a bond is broken in a silicon crystal a hole is created for each electron that leaves the bond.
e) When an electric field is applied to a semiconductor, a current flows due to conduction by both holes and electrons.
f) The impurity atoms in an n-type semiconductor have a valency of five.
g) An n-type semiconductor contains fixed positive donor ions and mobile electrons.
h) The conduction process in an p-type semiconductor is primarily due to holes.
i) The majority carriers in a p-type semiconductor are holes while the minority carriers are electrons.
j) Increasing the level of doping in a semiconductor increases the majority carrier concentration and decreases the minority carrier concentration.

2.

a) Each silicon atom has four valence electrons and forms covalent bonds with four neighbouring atoms.
b) Semiconductors in a highly purified state are called intrinsic semiconductors.
c) Polycrystalline semiconductor materials are used in the fabrication of some electronic devices, for example thermistors.
d) Silicon belongs to group IV of the periodic classification and crystallizes in the diamond structure.

e) When an electric field is applied, a current flows due to conduction by both holes and electrons.
f) At higher temperatures, electrons are elevated into the conduction band, leaving holes in the valence band.
g) The ionization of donor atoms results in fixed positive donor ions and mobile electrons.
h) At a given temperature, the product of the concentrations of the electrons and holes remains constant.
i) The introduction of impurities into a semiconductor is very often known as doping.
j) When an aluminium atom acquires an electron in this way, it becomes a negative ion.

3.

a) Many types of semiconductor materials are used in the fabrication of electronic devices. Some of these are elements like silicon and germanium, while others are intermetallic compounds like GaAs and InSb.
b) Silicon belongs to group IV of the periodic classification and crystallizes in the diamond structure. Each atom has four valence electrons and forms covalent bonds with neighbouring atoms.
c) Impurity atoms which when ionized give rise to fixed positive ions and mobile electrons are called donor atoms. Semiconductors with these properties are called n-type semiconductors.
d) The introduction of impurities into semiconductors is very often known as doping. Increasing the level of doping increases the concentration of majority carriers and decreases the concentration of minority carriers.
e) The conduction process in an n-type semiconductor is mainly due to electrons. For this reason, electrons are called majority carriers and holes minority carriers.

Exercises III

1.

a) A p-n junction can be formed by introducing donor impurities into one part of a crystal and acceptor impurities into the other part of a crystal.
b) When an external voltage is applied across the contacts of a p-n junction diode, the potential barrier is increased or decreased.
c) This is because the concentration of holes on one side of the junction is higher than on the other side. For electrons a similar situation exists with the sides being reversed.
d) The depletion region on either side of the junction is created because electrons and holes in the vicinity of the junction combine with each other.
e) The charge on the p-side of the junction is negative.
f) The potential barrier opposes further diffusion.
g) The positive terminal of an external battery is connected to the p-side of a junction diode for forward bias.
h) The width of the depletion layer is increased when a reverse bias is applied.
i) The height of the potential barrier is decreased when a forward bias is applied.
j) The total current is the sum of the electron and hole currents.

2.

a) A rectifier allows a large current to flow in the forward direction, and a small current to flow in the reverse direction.
b) There is a high concentration of holes on the p-side of the junction, and this causes them to diffuse across the junction.
c) The region on either side of the junction is depleted of electrons and holes, and is called a depletion or space charge region.
d) The charges on either side of the junction form a potential barrier which opposes further diffusion of electrons and holes.
e) Reverse bias increases the height of the potential barrier across the junction.
f) Forward bias decreases the width of the depletion layer.

g) Holes crossing the junction from the left are equivalent to electrons crossing from the right.
h) Under reverse bias conditions, the negative terminal of the batterry is connected to the p-side of the junction.
i) The potential barrier formed across the junction opposes the further diffusion of electrons and holes.
j) The electrons and holes combine with each other leaving negative acceptor ions on the left of the junction.

3.

a) A p-n junction can be formed in a semiconductor crystal by introducing donor impurities into one side of the crystal and acceptor impurities into the other side of the crystal.
b) A junction diode is a device with external metal contacts attached to the ends of a semiconductor crystal with a sharp p-n junction inside it. Such a device behaves like a rectifier, allowing a large current to flow in one direction and a negligible current to flow in the other direction.
c) Initially there is a high concentration of holes on one side of the junction, and a high concentration of electrons on the other side of the junction. This density gradient causes the holes and electrons to cross the junction and combine with each other.
d) Reverse bias increases the magnitude of the potential barrier across the junction and also the width of the depletion region. On the other hand, forward bias decreases the magnitude of the potential barrier, and the width of the depletion region
e) When a forward bias is applied, the equilibrium which was established between the diffusion of majority carriers and the opposing influence of the potential barrier is disturbed.

Exercises IV

1.

a) The junctions that exist between a metal and a semiconductor can be ohmic junctions or rectifying junctions.
b) The external voltage-current characteristic of a Schottky diode is similar to that of a junction diode.
c) A rectifying contact can be formed by depositing aluminium directly on n-type silicon, while an ohmic contact can be formed by introducing a heavily doped n+ layer between the aluminium and the n-type silicon.
d) The Schottky diode is a majority carrier device while the junction diode is a minority carrier device.
e) The junction diode has a larger delay time because the minority carriers stored near the junction must be removed before the diode can be switched from the forward to the reverse direction.
f) Metal-semiconductor point contact diodes can be used at microwave frequencies.
g) The width of the depletion layer is increased when the reverse voltage is increased.
h) Zener breakdown occurs when a strong electric field exists at a junction.
i) Varactor diodes can be used in automatic frequency control circuits at VHF and UHF frequencies.
j) The voltage-current characteristic of a tunnel diode shows a negative resistance region in the forward direction and a large current flow in the reverse direction.

2.

a) The junction formed between a metal and a semiconductor may be ohmic or rectifying.
b) The junction in a metal-semiconductor diode is a rectifying junction.
c) An n-type semiconductor is used with aluminium as the metal to form a Schottky diode.
d) In the forward direction, electrons from the semiconductor flow into the metal.

e) The Schottky diode is a majority carrier device as compared with the junction diode which is a minority carrier device.
f) The storage time for a Schottky diode is negligible because the current flow is mainly due to majority carriers.
g) If a p-n junction is reverse biased, electrons and holes move away from the junction.
h) The width of the depletion layer increases with increasing reverse voltage.
i) The process known as zener breakdown occurs when a strong electric field exists at a junction.
j) A tunnel diode shows a negative resistance region in the forward direction.

3.

a) The junction formed between a metal and a semiconductor can be ohmic or rectifying.The construction of metal-semiconductor diodes is only possible if rectifying contacts are used at the junction.
b) In a junction diode delay occurs when switching it from the forward to the reverse direction, because minority carriers stored near the junction must first be removed. The time taken to remove the stored minority carriers is called the storage time.
c) When a p-n junction diode is reverse biased, electrons and holes do not cross the junction but move away from the junction. Increasing the reverse voltage increases the width of the depletion layer.
d) When the reverse voltage applied to a junction diode is increased beyond a certain value, the current suddenly increases and the diode is said to be in the breakdown region. Two mechanisms of breakdown are recognized, avalanche breakdown and zener breakdown.
e) The width of the depletion layer in a junction diode depends on the impurity concentration. In a tunnel diode, the impurity concentration is increased until it is about 1/1000.

Exercises V

1.

a) The three semiconductor regions in a transistor are called emitter, base, and collector. The central region is narrow in width and has a lower impurity concentration than the other regions.
b) In a symmetrical transistor, the heights of the barriers at the two junctions are equal.
c) The emitter-base junction is forward biased and the collector based junction is reverse biased.
d) The height of the potential barrier at the first junction is reduced and the height at the second junction increased.
e) The main physical process which gives rise to the collector current I_C in an pnp transistor is the flow of holes from the emitter to the collector.
f) The base current in a pnp transistor is due to the flow of electrons from the base to the emitter and also due to holes which come from the emitter combining with electrons in the base region.
g) $I_E = I_B + I_C$.
h) The current I_{C0} is the reverse saturation current of the reverse biased collector-base junction.
i) The relationship is: $\beta = \alpha/(1-\alpha)$.
j) The large signal current gain refers to the ratio of the total currents, while the small signal current gain is the ratio of small increments in the currents.

2.

a) The central region of a transistor has a narrower width and a much lower conductivity than the other two regions.
b) When a transistor is on open circuit, there is no net flow of electrons or holes across the junctions.
c) When a transistor is used as an amplifier, the first junction is forward biased and the second junction is reverse biased.
d) Holes in the emitter region of a transistor move across the first junction into the base.

e) When holes reach the second junction the field is in such a direction as to accelerate them across the junction.
f) The flow of electrons out of the base constitutes part of the base current in a pnp transistor.
g) Most of the holes in a pnp transistor flow from the emitter to the collector without combining with electrons in the base region.
h) The current I_{C0} is the reverse saturation current of the reverse biased collector-base junction.
i) The ratio of the collector current to the emitter current is represented by the large signal current gain parameter α.
j) The ratios of the currents are important parameters in studying the behaviour of a transistor.

3.

a) When no external voltages are applied, holes or electrons do not cross the two p-n junctions of a transistor. For a symmetrical transistor the heights of the potential barriers at the junctions are equal.
b) When a transistor is used as an amplifier the emitter-base junction is forward biased, while the collector-base junction is reverse biased. The potential barrier at the first junction is lowered, while the barrier at the second junction is increased in height.
c) Holes in the emitter region move across the first junction into the base region, because the forward bias lowers the potential barrier across the junction below the equilibrium value.
d) Most of the holes drift through the narrow base region without combining with electrons in the base, because the base has a low electron concentration. When the holes reach the second junction, the field is in such a direction as to accelerate them across the junction.
e) Transistors are designed so as to make the base current small in comparison to the emitter current. This is achieved by having a base region of narrow width and low impurity concentration. Most of the holes flow from the emitter to the collector without combining with the electrons in the base region.

Exercises VI

1.

a) A transistor amplifier can have voltage gain, current gain, and power gain.
b) In a common base circuit, the base terminal is common to both input and output circuits. It has a high voltage gain and a high power gain, but has a current gain of slightly less than unity.
c) The input voltage is applied between the emitter and the base.
d) The output voltage is developed across the load resistor.
e) The voltage gain is defined as the ratio of the voltage across the load to the input voltage.
f) The common base amplifier has a high voltage gain because the load resistor can be made much larger than the input impedance.
g) The load is connected between the emitter and the common terminal.
h) It is useful for the matching of low impedance loads.
i) The voltage gain is slightly less than unity.
j) The common emitter circuit has a higher power gain than the other circuits.

2.

a) The transistor is widely used as an amplifying device.
b) In the common base circuit, the base terminal is common to both input and output circuits.
c) In a common base circuit, the input voltage is applied between the emitter and the base.
d) The output voltage is developed across the load resistor.
e) The common base circuit has a high voltage gain and a high power gain, but has a current gain of slightly less than unity.
f) In a common collector circuit, the load resistor is connected between the emitter and the common terminal.
g) A common collector circuit is used as an impedance transformer for matching low impedance loads.
h) A common base circuit has a low input impedance and a high output impedance.

i) There are three possible ways in which a transistor can be conmnected in an amplifier circuit.
j) The popularity of the common emitter circuit is due to the fact that its power gain is higher than for the other circuits.

3.

a) There are three possible ways in which a transistor can be connected in an amplifier circuit. The circuits are called common base, common emitter, and common collector circuits, depending on whether the base, emitter, or collector terminals of the transistor are common to both input and output circuits.
b) In a common collector or emitter follower circuit, the load resistor is connected between the emitter and the common terminal. The voltage gain of this circuit is slightly less than unity, but it has high values of current and power gain.
c) The common collector has a high input impedance and a low output impedance. Such a circuit is often used as an impedance transformer for the matching of low impedance loads.
d) The input voltage is applied between the emitter and the base, while the output voltage is developed across the load resistor. A small change in input voltage produces a large change in emitter current.
e) The common emitter circuit is the most frequently used of the three circuits. The popularity of this circuit is due to the fact that its power gain is higher than for the other two circuits.

Exercises VII

1.

a) The D.C operating conditions have to be optimized so that the transistor will operate as linearly as possible.
b) The operating conditions are optimized by using a graphical method.
c) The output characteristics consist of a family of curves which show how I_C varies when V_{CE} is changed.
d) In the active region the emitter junction is forward biased and the collector junction reverse biased.
e) An appropriate hyperbola drawn on the output characteristics corresponds to the maximum power rating of the transistor.
f) The transistor must remain in the active region if it is to operate with maximum linearity.
g) The maximum values of I_C and V_{CE} should not be exceeded.
h) The values of the D.C supply voltage, load resistor, operating point, and bias resistor have to be chosen in successive steps.
i) A suitable value for the load resistor is found by drawing an appropriate load line on the output characteristics.
j) A self bias circuit has better stability than the fixed bias circuit because it compensates for changes in the operating point due to various causes.

2.

a) When a transistor is used in a circuit, the D.C operating conditions have to be first optimized.
b) This is done by a graphical method which uses two families of characteristic curves.
c) In the active region the emitter junction is forward biased, and the collector junction is reverse biased.
d) To obtain maximum linearity of operation the transistor must be kept within the active region.
e) It is normal to use a single D.C voltage supply instead of two batteries to obtain the biasing voltages for a transistor circuit.
f) The dotted hyperbola drawn on the output characteristics is the locus corresponding to the maximum power rating of the transistor.

g) To find a suitable value of load resistor, a straight line called a load line is drawn on the output characteristics.
h) The output characteristics show how I_C varies when V_{CE} is changed.
i) Other limitations are that the maximum values of I_C and V_{CE} must not be exceeded.
j) If the temperature changes or if the transistor is changed, the operating point changes.

3.

a) When a transistor is used in a circuit, the D.C operating conditions have to be optimized first, so that the transistor will operate as linearly as possible.The largest possible output voltage should be produced with the minimum of distortion.
b) To obtain maximum linearity of operation, the transistor must be kept within the active region and below the maximum power rating curve.
c) Normally a single voltage is used instead of two batteries to obtain the biasing voltages for a transistor circuit. The simplest type of fixed bias circuit is shown in the Fig 7.3.
d) A suitable supply voltage can be chosen by using the manufacturer's transistor data sheets. To find a suitable value of load resistor, a load line is drawn on the output characteristics.
e) The fixed bias circuit has poor stability. Changes in temperature cause changes in the current I_{C0}.The variations in the values of β for different transistors of the same type can also be large.

Exercises VIII

1.

a) The first step in designing a transistor amplifier stage is the establishment of the D.C operating conditions.
b) The quantitative analysis of a transistor circuit can be carried out by using equivalent circuits.
c) The small signal parameters are usually obtained from the manufacturer's data sheets. They can also be obtained experimentally by the user.
d) Two voltages and two currents are required for studying the behaviour of a four terminal network.
e) The hybrid parameters can be considered to be constants when the circuit is linear.
f) The word hybrid is used to describe the parameters because they are dimensionally different from each other.
g) The subscripts e, b, c, are used to indicate whether the circuit is common emitter, common base, or common collector.
h) The parameter h_{21} is the short circuit current gain parameter, and h_{22} corresponds to the output conductance with input short-circuited.
i) The letter i corresponds to the input, the letter f to forward transfer, the letter o to output, and the letter r to reverse transfer.
j) The parameter h_{re} can usually be omitted because it is very small.

2.

a) The first step in designing a transistor amplifier stage is to establish the D.C operating conditions by using a graphical method.
b) When this has been done, small signals applied to the input of a transistor circuit are amplified with reasonable linearity.
c) The analysis of a transistor circuit can be carried out by using equivalent circuits which represent the behaviour of the transistor.
d) The small signal parameters used in the equivalent circuits are usually supplied by the manufacturer.
e) The most commonly used equivalent circuit is the hybrid equivalent circuit, and its parameters can be determined by experiment.

f) The transistor circuit is replaced by an equivalent circuit.
g) If the circuit is linear, then the hybrid parameters are constants.
h) The parameter h_{11} is defined as the input resistance with output short-circuited.
i) For transistor circuits, the letters e, b, c, are added to indicate the kind of circuit.
j) In the simplified hybrid equivalent circuit, the parameter h_{12} can be omitted because it is small.

3.

a) The first step in designing an amplifier stage is to establish the D.C operating conditions graphically. After this has been done, small signals which are applied to the input are amplified with reasonable linearity.
b) A transistor circuit can be replaced by an equivalent four terminal network. The behaviour of such a network can be studied in terms of two voltages and two currents.
c) The small signal parameters used in the equivalent circuits are usually supplied by the manufacturer. They can also be determined experimentally for a given transistor by the user.
d) In order to obtain maximum linearity, the transistor must operate in the region below the curve corresponding to the maximum power rating of the transistor. The maximum values of I_C and V_{CE} given by the manufacturer should not be exceeded.
e) The hybrid parameter h_{11} is defined as the input resistance with the output short-circuited, while the parameter h_{21} gives the ratio of the output to the input current and is called the short circuit current gain.

Exercises IX

1.

a) A wideband amplifier is an amplifier that can amplify over a wide band of frequencies, from a few hertz to several megahertz.
b) The gain of the amplifier should be independent of the frequency and the phase shift should be proportional to the frequency.
c) A nonsinusoidal signal is composed of many frequencies.
d) In the midfrequency region, the gain and time delay remain reasonably constant.
e) The decrease in gain with decreasing frequency is due to the effect of the external coupling capacitor and the emitter resistor bypass capacitor.
f) The response of an amplifier in the high frequency region is similar to that of a lowpass filter.
g) The bandwidth of an amplifier is the frequency range between the upper and lower 3 dB frequencies.
h) The rise time of an amplifier is the time taken for the voltage to rise from one-tenth to nine-tenth of its final value.
i) The low frequency response of an amplifier determines the sag in the flat portion of the output waveform.
j) The 3 dB frequencies are the frequencies at which the gain of the amplifier is 0.707 times the midfrequency gain.

2.

a) A typical example of a wideband amplifier is the video amplifier in a television receiver.
b) The gain of an ideal amplifier should be independent of the frequency, and the phase angle should be proportional to the frequency.
c) The gain and phase shift of an amplifier vary with the frequency of the input signal.
d) In the midfrequency region, the gain and time delay remain reasonably constant.
e) In the low frequency region, the gain decreases with decreasing frequency.

f) The response of an amplifier in the high frequency region is similar to that of a lowpass filter.
g) The low frequency response of an amplifier is closely related to the sag in the flat portion of the output waveform.
h) The rise time of an amplifier is the time taken for the voltage to rise from ten percent to ninety percent of its final value.
i) The rise time is a measure of how fast an amplifier can respond to a sudden change in input voltage.
j) Another way of assessing the behaviour of an amplifier is to consider its response to a step waveform.

3.

a) Wideband amplifiers can amplify over a wide band of frequencies, from a few hertz to a few megahertz. A typical example of such an amplifier is the video amplifier in a television receiver.
b) The gain of the amplifier should be independent of the frequency, and the phase shift should be proportional to the frequency. This means that the time delay should be the same for all the frequency components of a signal.
c) In the low frequency region the gain decreases with decreasing frequency due to the effect of the coupling capacitor and the emitter resistor bypass capacitor.
d) The circuit of a typical RC coupled amplifier is shown in Fig 9.1. The amplitude-frequency response characteristic can be divided into three regions, midfrequency, low frequency, and high frequency.
e) The time taken for the voltage to rise from one-tenth to nine-tenth of its final value is known as its rise time. The rise time is a measure of how fast an amplifier can respond to a sudden change in input voltage.

Exercises X

1.

a) The response of a transistor to input voltages and currents is not instantaneous because it takes time for the charge carriers to flow through the transistor.
b) The short circuit current gain decreases at high frequencies, because the transistor takes time to respond to an input signal.
c) The hybrid-π circuit is used to represent the behaviour of a transistor at high frequencies.
d) The values of the resistances can be obtained from the low frequency parameters.
e) They do not vary with frequency.
f) The range of frequencies up to the upper 3 dB frequency is called the bandwidth.
g) At the frequency f_T, the short circuit common emitter current gain becomes unity.
h) The frequency f_T is equal to the product of h_{fe} and f_β.
i) C_C is the measured value of the collector to base output capacitance with open input.
j) C_e is the sum of the emitter diffusion capacitance C_{De} and the emitter junction capacitance C_{Te}.

2.

a) The flow of charges takes time, and the response of the transistor is not instantaneous.
b) This results in a gradual decrease in the value of the short circuit current gain at high frequencies.
c) A circuit that has been widely used to represent the behaviour of a transistor at high frequencies is the hybrid-π circuit.
d) The resistances and capacitances used in this circuit are assumed to be independent of frequency.
e) The values of the resistances and conductances used in a hybrid-π circuit can be obtained from the low frequency h-parameters.

f) The range of frequencies up to the upper 3 dB frequency is known as the bandwidth of the circuit.
g) At the frequency f_T the magnitude of the short circuit current gain becomes unity.
h) The frequency f_T is the product of the short circuit current gain and the upper 3 dB frequency.
i) The currents flowing through a transistor result from the flow of charge carriers between the different regions of the transistor.
j) The capacitance C_C is the measured value of the collector to base output capacitance with open input.

3.

a) The currents flowing through a transistor result from the diffusion of charge carriers between the different regions of the transistor. As the diffusion process takes time, the response of the transistor to changes in the input voltages and currents is not instantaneous.
b) The magnitude of the common emitter short circuit current gain decreases with frequency. The equivalent circuit used to represent the high frequency behaviour of a transistor must therefore be different from the circuit used at low frequencies.
c) An equivalent circuit that is widely used to represent the behaviour of a transistor at high frequencies is the hybrid-π circuit. The resistances and capacitances used in this circuit are independent of frequency.
d) The values of the resistances and conductances can be obtained from the low frequency h-parameters. The capacitance C_C is the measured value of the collector to base output capacitance with open input.
e) An important parameter is the transit frequency f_T, which is the frequency at which the short circuit common emitter current gain becomes unity. A graph showing the variation of the short circuit current gain with frequency is shown in the Fig 10.4.

Exercises XI

1.

a) Negative feedback can have the effect of modifying and improving the characteristics of an amplifier.
b) Negative feedback tends to oppose an increase in the magnitude of the output signal.
c) Reasons for variations in gain include large variations in transistor parameters as well as parameter variations due to temperature changes.
d) Negative feedback can reduce hum, noise, distortion, etc in an amplifier.
e) The input and output impedances can be increased or decreased by the application of negative feedback.
f) The bandwidth of an amplifier is increased by the application of negative feedback.
g) Negative feedback reduces the midfrequency gain of an amplifier.
h) Negative feedback may produce instability in an amplifier and cause it to oscillate.
i) The distortion is reduced by a factor of $1 / (1 + A\beta)$.
j) The relationship between the feedback factor and the gain with feedback is $A_f = A / (1 + A\beta)$.

2.

a) The characteristics of an amplifier can be modified and improved by the use of negative feedback.
b) The gain of an amplifier can be stabilized by the use of negative feedback.
c) The input and output impedances of an amplifier can be increased or decreased by the application of negative feedback.
d) The application of negative feedback increases the bandwidth at the expense of gain.
e) The midfrequency gain of an amplifier is reduced by the application of negative feedback.
f) The application of negative feedback may result in the amplifier becoming unstable and breaking into oscillations.

g) The gain of an amplifier usually decreases at high frequencies.
h) The phase shift of the amplified signal varies with frequency.
i) Variations in temperature, and changes in the values of the transistor parameters can cause large changes in the gain of an amplifier.
j) Distortion which is produced in an amplifier can be reduced by negative feedback.

3.

a) The characteristics of an amplifier can be improved by the use of negative feedback. In this process, a part of the output signal is combined with the input signal.
b) Changes in temperature and other factors like changes in transistor parameters can cause large changes in the gain of an amplifier. Transistors of the same type may have parameters which vary by as much as 50 %.
c) The application of negative feedback stabilizes the amplifier and reduces variations in gain due to any cause. Distortion, noise, and hum, are also reduced.
d) The input and output impedances of an amplifier can be increased or decreased by the application of negative feedback. The bandwidth of the amplifier is increased at the expense of gain.
e) The application of negative feedback may cause the amplifier to become unstable. This is due to the fact that the phase shift of the output signal varies with frequency.

Exercises XII

1.

a) IC operational amplifiers have the advantages of versatility, reliability, small size, low cost, and the ability to amplify D.C signals.
b) Typical voltage gains are of the order of 2×10^5, and bandwidths are of the order of a few megahertz.
c) The two input terminals are the inverting input and the noninverting input.
d) A single-ended operational amplifier may be constructed by earthing one of the input terminals.
e) Feedback is applied to an inverting amplifier by using two external impedances Z and Z' as shown in Fig 6.2
f) The gain of an operational amplifier with feedback depends only on the values of Z and Z'.
g) In a noninverting amplifier, the input and output voltages are in phase with each other.
h) Drift in a D.C amplifier can be reduced by using a balanced circuit.
i) The output voltage is proportional to the voltage difference between the two input voltages.
j) Imbalance is caused by the fact that the two input transistors are not perfectly matched.

2.

a) The integrated circuit operational amplifier has been widely used as a building block in analog circuits.
b) An operational amplifier has two input terminals and one output terminal.
c) A single-ended amplifier can be constructed by earthing one of the input terminals.
d) If the input impedance of the amplifier is assumed to be infinity, no current flows into the amplifier input terminals.
e) Both input terminals are virtually at the same potential and the inverting terminal is at earth potential.

f) In a noninverting circuit the output voltage is in phase with the input voltage.
g) In a balanced circuit voltages in one part of the circuit are balanced by equal and opposite changes in another part of the circuit.
h) In a differential amplifier, the output voltage is proportional to the difference between the input voltages.
i) Voltage changes can occur as a result of temperature variations and are termed drift voltages.
j) The quality of a differential amplifier may be assessed by measuring its common mode rejection ratio.

3.

a) The integrated circuit operational amplifier is widely used as a building block in analog circuits. It has the advantages of versatility, low cost, high reliability, and the ability to amplify D.C signals.
b) The operational amplifier is a multistage direct coupled high gain amplifier, whose overall amplification can be controlled by the addition of negative feedback.
c) For practical purposes, one can assume that an operational amplifier has infinite voltage gain, infinite input impedance, zero output impedance, and a large bandwidth.
d) A good way of reducing amplifier drift is to use a balanced circuit, where the voltages in one part of the circuit are balanced by equal and opposite changes in another part of the circuit.
e) An ideal operational amplifier shows perfect balance. This means that the output voltage is zero when the voltages at both inputs are zero. Normally the transistors in the input stage are not perfectly matched and this causes some imbalance.

Exercises XIII

1.

a) The IC operational amplifier is used as a building block in many analog systems.
b) The term operational amplifier is used because circuits containing these amplifiers can perform mathematical operations.
c) A circuit that can perform mathematical operations can be built from an IC operational amplifier and a few discrete components.
d) A multiplier circuit can be constructed by using an inverting amplifier and two resistances.
e) A phase shifter can be constructed by using an inverting amplifier and two impedances. The impedances should be equal in magnitude, but have different phase angles.
f) An analog computer consists of a combination of circuits which can together perform certain mathematical operations like for example the solution of differential operations.
g) Resistors and capacitors are used together with an operational amplifier to form an active filter.
h) Inductances are expensive and tend to pick up hum from the mains.
i) A flat-topped bandpass filter can be constructed by connecting two active filters in cascade. The resonant frequencies are slightly staggered.
j) Some of the linear systems which use operational amplifiers are voltage amplifiers, voltage to current converters, and delay equalizers.

2.

a) The IC inverting amplifier has been used as a basic building block in the construction of analog systems.
b) Circuits containing operational amplifiers and a few discrete components can perform many mathematical operations.
c) A phase shifter can be constructed using the inverting amplifier as a basis.
d) In an integrator, a capacitance C is used for Z' and a resistance for Z.
e) In a differentiator, the output voltage is proportional to the time derivative of the input voltage.

f) Filters are often used for the purpose of restricting the frequency bandwidth of a circuit.
g) Passive resistors use a combination of inductances, capacitances, and resistances, to achieve the necessary frequency response.
h) Inductances are expensive and tend to pick up hum from the mains.
i) A flat-topped bandpass characteristic can be obtained by using active filters in cascade.
j) By choosing suitable circuits and component values, an analog computer can be used to solve differential equations.

3.

a) Circuits which contain operational amplifiers and a few discrete components can perform many mathematical operations, like addition, multiplication, differentiation, and integration.
b) In a multiplying circuit, the output voltage is k times the input voltage. Almost any value of k can be obtained by choosing suitable values of resistors for Z and Z'. Such a circuit can also be used as a divider.
c) An inverting amplifier can perform the mathematical operation of integration if a capacitance C is used for Z' and a resistance R for Z. The output voltage is equal to the integral of the input voltage.
d) Circuits like the differentiator, the integrator, the adder, etc. can be combined together to form an analog computer. Such a computer can be used to solve differential equations.
e) The use of inductances can be avoided by using active filters. It is possible to simulate the behaviour of an LCR filter by a circuit that contains only resistors, capacitors, and an operational amplifier.

Exercises XIV

1.

a) The operation of the junction transistor depends on the flow of both majority and minority carriers, while the operation of the field effect transistor depends only on the flow of majority carriers.
b) The two main types of field effect transistors are the junction field effect transistor and the metal oxide field effect transistor.
c) The FET is simpler to fabricate than the junction transistor. It occupies less space in an integrated circuit, it is less noisy, and has a high input impedance.
d) Its main disadvantage is that its gain-bandwidth product is smaller than that of the junction transistor.
e) The source and drain electrodes are ohmic metal contacts attached to the ends of a bar of silicon.
f) The flow of current can be controlled by applying a potential difference between the gate and the source
g) A reverse bias has to be applied to the p-n junctions.
h) In a depletion MOSFET, an impurity is permanently diffused into the region between the source and the drain. In the enhancement MOSFET; current flow takes place due to charge carriers which are attracted into the region between the source and the drain as a result of the application of a voltage between the source and the gate.
i) The depletion MOSFET can be operated in both enhancement and depletion modes.
j) This is because there are two opposing junctions p-n and n-p between the source and the drain.

2.

a) The operation of the junction transistor depends on the flow of both majority and minority charge carriers.
b) The field effect transistor is easier to construct and occupies less space in an integrated circuit.
c) A field effect transistor is constructed from a bar of silicon which has an ohmic metal contact at each end.

d) The main disadvantage of the field effect transistor is that it has a smaller gain-bandwidth product than the junction transistor.
e) Majority carriers enter the bar through the source and flow out through the drain.
f) In a field effect transistor the flow of majority carriers can be controlled by applying a voltage between the gate and the source.
g) In a depletion MOSFET an impurity of the same type as present in the source and drain is permanently diffused into the region between the source and drain.
h) In an enhancement MOSFET no drain current flows unless a suitable potential difference is applied between the source and the gate.
i) In an enhancement MOSFET, the flow of current is due to the induced channel produced by applying a potential difference between the gate and the source.
j) A depletion MOSFET can also be operated in the enhancement mode.

3.

a) The operation of the bipolar junction transistor depends on the flow of majority and minority carriers. The operation of the field effect transistor however depends only on the flow of majority carriers.
b) The field effect transistor has many advantages over the bipolar junction transistor. It is simpler to fabricate and occupies less space in an integrated circuit. Further it is less noisy, and has a higher input resistance.
c) The n-channel FET is constructed from a bar of n-type silicon which has an ohmic contact at each end. The contacts at the end of the silicon bar are called the source and the drain.
d) The flow of majority carriers from source to drain can be controlled by applying a potential difference betweeen the gate and the source. The gate therefore acts as a control electrode.
e) If a negative gate voltage is applied, positive charges are induced in the channel and this causes a depletion in the number of majority carriers. This is the reason for the use of the term depletion MOSFET.

Exercises XV

1.

a) The factor that is common to all oscillator circuits is positive feedback.
b) The amplifier produces a signal at the output without any input.
c) A frequency selective network has to be included in the feedback path.
d) The condition for oscillation is $1 - A\beta = 0$.
e) The oscillator circuits used at low frequencies are RC oscillator circuits.
f) A circuit that is very popular as the basis for a low frequency oscillator is the Wien bridge circuit.
g) LC oscillators are used at frequencies above 50 kHz.
h) When a piezoelectric crystal is mechanically stressed, a voltage is produced across two opposite faces of the crystal.
i) This is because the crystal behaves like an electrical resonant circuit of very high Q factor.
j) The output level is usually controlled by using a monitoring meter and an attenuator whose impedance is constant at all settings.

2.

a) The factor that is common to all oscillator circuits is positive feedback.
b) To ensure that oscillation occurs at a definite frequency it is necessary that a frequency selective network be included in the feedback path.
c) At low frequencies, RC circuits are more commonly used than LC circuits.
d) The inclusion of the bulb in the negative feedback circuit tends to stabilize the output of the oscillator.
e) At frequencies above 50 kHz, the LC circuit is the most commonly used frequency selective network.
f) Certain crystals, notably quartz and some ceramics, exhibit piezoelectric properties.
g) When a piezoelectric crystal is stressed mechanically, a voltage is produced between two opposite faces of the crystal.
h) A piezoelectric crystal vibrates mechanically when an A.C voltage is applied across two of its faces.

i) A quartz crystal can be used as the frequency selective network of an extremely stable fixed frequency oscillator.
j) The carrier waveform produced by a quartz oscillator in a TV transmitter is very different from the waveform produced by the time base generator in a TV receiver.

3.

a) A large variety of circuits and instruments have been developed for the generation of periodic and nonperiodic signals, as different types and shapes of signals are required for different purposes.
b) The carrier waveform generated by the quartz oscillator in a TV transmitter is very different from the waveform produced by the time-base generator in a TV receiver.
c) Certain crystals, notably quartz and some ceramics, exhibit piezoelectric properties. When the crystal is stressed mechanically, an electric voltage appears between the opposite faces of the crystal.
d) The crystal behaves like a resonant circuit of high Q factor and low damping. It can be used as the frequency selective network of an extremely stable fixed frequency oscillator.
e) A signal generator is an instrument which can produce a waveform of the desired shape and frequency, and whose output voltage or power can also be set to a definite value.

Exercises XVI

1.

a) Silicon controlled rectifiers have the advantage that they are able to control the power delivered to the load.
b) SCRs may be used in applications like motor speed control, lighting control, and electrical welding.
c) The layers of semiconductor material in a four layer diode are arranged in the order p-n-p-n.
d) The two stable states of a four layer diode are called the ON state and the OFF state.
e) When the applied voltage exceeds the break-over voltage V_{BO}, the diode switches from its OFF state to its ON state.
f) If the voltage is reduced, the SCR remains in the ON state until the current decreases to I_H. The current I_H and the corresponding voltage V_H are called the latching current and latching voltage respectively.
g) The gate acts as a control electrode and the gate-cathode currents control the anode to cathode break-over voltage.
h) Two SCR devices can be connected in inverse parallel to deliver full power.
i) The triac has the advantage over two inverse parallel connected SCRs in that it needs only one source of gate pulses.
j) The conduction angle can be controlled by varying the phase of the gate pulses relative to the A.C supply.

2.

a) The use of silicon controlled rectifiers has made electrical power control an efficient and inexpensive process.
b) A four layer diode is composed of four layers of silicon arranged in the order p-n-p-n.
c) A silicon four layer diode has two stable states.
d) If the forward voltage is increased beyond V_{BO}, the diode switches from its OFF state to its ON state.
e) The SCR is a rectifier which has the added ability to control the power delivered to the load.

f) The gate acts as a control electrode and currents flowing in the gate-cathode circuit may be used to control the anode to cathode break-over voltage.
g) The SCR remains OFF and no current flows until it is turned ON by the application of a gate current pulse.
h) The SCR is a half wave device and is only able to deliver half power even at full conduction.
i) The triac may be considered to be composed of two inverse paralleled thyristors which are both controlled by a single gate electrode.
j) The conduction angle is controlled by the phase of the gate pulses relative to the A.C supply.

3.

a) Solid state devices play an important role in the electrical power control today. Many applications require a variable but controlled amount of current.
b) Among these applications are motor speed control, electrical welding, and lighting control. The use of silicon controlled rectifiers has made electrical power control an efficient and inexpensive process.
c) When the anode of a thyristor is made positive with respect to the cathode, the junctions J_1 and J_3 are forward biased while J_2 is reverse biased. With forward bias, the diode has two stable states, one a high resistance state called the OFF state, and another a very low resistance state called the ON state.
d) As its name implies, the SCR is a rectifier which has the added ability of controlling the power delivered to the load. The SCR has a structure similar to the four layer diode with an added electrode called the gate as shown in the figures.
e) The most useful device for A.C power control is the triac or bidirectional thyristor. This can be considered to be composed of two inverse-paralleled thyristors which are both controlled by the same gate.

Exercises XVII

1.

a) Some of the systems based on digital circuits are computers, data processing systems, and digital communication systems.
b) Digital systems have many advantages over analog systems and are used for the processing of signals which have been converted from analog into digital form.
c) A digital system uses binary devices which have only two states.
d) A transistor can be used as a binary device.
e) A digital system functions by the repetition of a few basic operations.
f) The circuits used to perform the basic operations in digital circuits are called gates.
g) The output of a two input XOR gate is in the 1 state if and only if one input is in the 1 state. In an OR gate, the output is in the 1 state if one or both inputs are in the 1 state.
h) The output of an AND gate is in the 1 state, if and only if all the inputs are in the 1 state.
i) The XOR gate can be constructed by using a number of gates to implement the expression $Y = (A + B)(\overline{AB})$.
j) An OR gate can be converted into an AND gate by inverting all the inputs and also the output.

2.

a) Digital circuits form the basis of a large number of electronic systems like computers and data processing systems.
b) A digital system uses binary devices which can have only two states.
c) A digital system functions by the repetition of a few basic operations.
d) The circuits used to perform these operations are called logic circuits.
e) An OR gate has two or more inputs and one output.
f) An electronic device which can operate in a binary fashion is a transistor which is allowed to be at saturation or at cut-off.
g) The EXCLUSIVE OR gate recognizes only an odd number of 1s at the input.

h) De Morgan's laws show that certain types of circuits are logically equivalent.
i) An OR gate can be converted into an AND gate by inverting all the inputs and also the output.
j) The type of algebra that is relevant to digital systems is called Boolean algebra.

3.

a) Digital circuits form the basis of a large number of electronic systems like computers, data processing systems, and digital communication systems.
b) A digital system however complicated functions by the repetition of a few basic operations. The circuits used to perform these operations are called logic gates.
c) The output of an EXCLUSIVE OR gate with two inputs is in the 1 state only if one of the inputs is in the 1 state. This gate can be constructed from simpler gates in many ways. One way of doing this is is shown in Fig 17.7.
d) These gates are the building blocks from which more complicated digital systems are built. The type of algebra that is relevant to digital systems is called Boolean algebra.
e) An OR gate has two or more inputs and one output. Its operation is in accordance with the definition: The output of an OR gate is in the 1 state, if one or more inputs are in the 1 state.

Exercises XVIII

1.

a) Four of the available types of logic families are: Diode Transistor Logic, Transistor Transistor Logic, Emitter Coupled Logic, and Complimentary MOS Logic.
b) Among the factors to be considered are speed, power dissipation, cost, fanout, and availability.
c) The advantages possessed by CMOS gates are their low power dissipation and the fact that they occupy less space in an integrated circuit.
d) TTL gates have the advantages of speed, availability, and low output impedance.
e) Some of the TTL families available in addition to standard TTL are, high speed TTL, low power TTL, and Schottky TTL.
f) Fanout is defined as the number of loads that can be driven from a single source.
g) A CMOS gate is composed of both n-channel and p-channel gates.
h) A CMOS inverter circuit consists of a p-channel transistor and an n-channel transistor connected in series.
i) Power is mainly dissipated when a CMOS gate changes state.
j) The most popular series of TTL gates is the 7400 series and the equivalent CMOS series is the MM74C00 series.

2.

a) Many types of electronic circuits have been used to implement logic gates.
b) Fanout is defined as the number of loads that can be driven from a single source.
c) The integrated circuits are classified into families depending on the type of circuits used for the gates.
d) The speed of a logic family is measured in terms of the propagation delay time of its basic NAND gate.
e) CMOS gates are popular because of their low power dissipation and the small space that they occupy in an integrated circuit.

f) Power is dissipated in a CMOS gate only when it switches from one state to another.
g) Standard TTL chips are available in a larger number of circuits than other families of circuits.
h) The power consumed in a CMOS gate is proportional to the frequency at which the gate is switched.
i) CMOS gates are slower in action than TTL gates and are not available in such a large variety of circuits as TTL gates.
j) A useful series of CMOS gates is the MM74C00 series in which each IC has the same function as the equivalent TTL in the 7400 gate series.

3.

a) Many types of electronic circuits have been used to implement logic gates. These gates are invariably manufactured in the form of integrated circuits.
b) Standard TTL ICs are available in a larger variety of circuits than chips of other logic families. They are faster than CMOS circuits but slower than ECL circuits.
c) CMOS ICs are extremely popular because they have low power dissipation and because of the small area which they occupy in an integrated circuit.
d) Power is dissipated in a CMOS gate only when it switches from one state to another. The power dissipation is therefore proportional to the frequency at which the gate is switched.
e) The logic family selected depends on the application for which it is intended. Some of the factors which have to be considered in selecting a logic family are speed, cost, and power dissipation.

Exercises XIX

1.

a) A digital system stores binary information in its memory elements.
b) A flip-flop is a clock controlled electronic device that has two states.
c) An SR latch can be constructed from two cross-coupled NAND gates.
d) The output of an SR flip-flop is unpredictable when both inputs are 1.
e) The ambiguous state which occurs in the SR flip-flop does not occur in the JK flip-flop.
f) Clock control is necessary in digital systems because data has to be entered at definite times.
g) There is no change in the output of a clocked SR flip-flop between clock pulses.
h) The preset and clear controls are used to assign a definite state to a flip-flop before starting an operation.
i) The master is enabled during a clock pulse, while the slave is enabled during the time between clock pulses.
j) In a toggle type flip-flop J = K = 1, and the output changes with each clock pulse.

2.

a) A digital system needs memory elements which are devices that can store binary information.
b) A flip-flop remains indefinitely in one state until it is triggered into the other state.
c) In digital systems it is usually necessary that data be entered in at a definite time.
d) A latch that can change state only during a clock pulse is called a flip-flop.
e) When the power is switched on, a flip-flop may be in an unknown state.
f) It is usually necessary to assign a definite state to a flip-flop before starting an operation.
g) The master is enabled for the duration of a clock pulse and the slave is disabled.

h) Clock pulses which are applied to the master are inverted before being applied to the slave.
i) The D-type flip-flop is a modified form of the JK flip-flop in which an inverter is included in the input.
j) The D-type flip-flop functions as a delay device because the input at D is transferred to the output at the next pulse.

3.

a) A digital system needs memory elements that can store binary information. One of the simplest types of memory elements is the bistable multivibrator, also called a flip-flop.
b) A flip-flop is an electronic device that has two stable states. It remains indefinitely in one of these states until it is triggered into the other state.
c) When the power is switched on a flip-flop may be in an unknown state, and because of this it is usually necessary to assign a definite state to flip-flops before starting an operation. This may be done by using the preset and clear inputs of the flip-flop.
d) Clock pulses which are applied to the master flip-flop are inverted before being applied to the slave flip-flop. The master is enabled for the time duration of the clock pulse, while the slave is disabled and cannot change state for the duration of the clock pulse.
e) The D-type flip-flop is a modified form of the JK flip-flop in which an inverter is included in the input as shown in the corresponding diagram.This ensures that K is always the complement of J and that an ambiguous state is avoided.

Exercises XX

1.

a) A flip-flop can store a single binary digit.
b) A shift register is constructed by combining a number of flip-flops.
c) The name shift register originates from the fact that this device takes in one new digit for each clock pulse, while at the same time shifting existing digits by one stage.
d) Some ways of entering and extracting data are, serial input serial output, parallel-input parallel-output, and parallel-input serial-output.
e) Data bits can be simultaneously entered into a shift register by using the preset inputs.
f) Data bits can be removed simultaneously from the Q outputs of a shift register at a selected time.
g) A right-left shift register is a device in which data bits can be shifted to the left or to the right.
h) If a clock pulse is applied to a shift register, each bit is moved to the next higher significant place, which is equivalent to a multiplication by two.
i) This is because an input pulse train entering an n-stage shift register appears at the output after a time (n – 1) T, where T is the period of the clock pulse.
j) A ring counter is constructed by connecting the last Q output of the shift register to the input terminal.

2.

a) The storage of a number of binary digits can be done by combining a number of flip-flops in a device called a shift register.
b) The shift register takes in one new digit for each clock pulse while moving the existing digits by one stage.
c) The flip-flops in a shift register are first cleared using the Cr and Pr inputs so that all the outputs are zero.
d) Each digit is now available on a separate output and may be read simultaneously to obtain a parallel output.

e) Shift registers are fitted with gates which allow the bits to be shifted to the left or to the right.
f) A left-right shift register can be used to perform multiplication or division.
g) If a clock pulse is applied to a left shift register, each bit is moved to the next higher significant place.
h) An input train of pulses entering an n-stage shift register appears as a pulse train after a time (n – 1) T.
i) A ring counter can be constructed by connecting the last output terminal of a shift register to the input terminal.
j) One application for a ring counter is its use as a substitute for the distributor in an automobile engine.

3.

a) A flip-flop can store a single binary digit. The storage of a number of binary digits can be done by combining several flip-flops to form a shift register.
b) The name shift register originates from the fact that this device takes in one new binary digit for each clock pulse while shifting the existing binary digits by one stage to make room for the new bit.
c) Shift registers can be fitted with gates which allow the data bits to be shifted to the left or to the right. A left-right shift register can be used to perform multiplication or division by two.
d) A shift register may be used as a time delay device. If an input train of data pulses enters an n-stage shift register, then it appears as a pulse train at the output after a time (n – 1) T, where T is the clock period.
e) One application of a ring counter is its use as a substitute for the distributor in a car engine. A car with a four cylinder engine would need a ring counter with four flip-flops.

Exercises XXI

1.

a) The availability of counters in IC form has made electronic counting a reliable and inexpensive process.
b) The output of binary counters can be converted into decimal form by using suitable conversion circuits.
c) The basic element of the binary counter is the master-slave flip flop set to toggle on each clock pulse.
d) The ripple counter consists of a series of flip-flops, the output Q of each flip-flop being connected to the clock input of the next one.
e) The master changes state when the pulse at its input changes from 0 to 1. The slave changes state when the pulse falls from 1 to 0.
f) A ripple counter becomes ineffective when the total counter delay time is longer than the interval between pulses.
g) In a synchronous counter, all the flip-flops are clocked synchronously or simultaneously.
h) The rate of counting of a synchronous counter can be twice as high as that of a ripple counter.
i) Each stage of a synchronous counter toggles only on clock pulses that occur when the outputs of all less significant stages are 1.
j) An asynchronous counter is called a ripple counter because each stage responds only after the previous stage has completed its transition.

2.

a) The availability of counters in IC form has made electronic counting a reliable and inexpensive process.
b) The binary output of a counter can be converted into decimal form by the use of suitable converting circuits.
c) The basic element of a binary counter is the master-slave flip-flop set to toggle on each clock pulse.
d) A ripple counter consists of a series of flip-flops, the output Q of each flip-flop being connected to the clock input of the next one.
e) In a toggle type flip flop, the master changes state when the pulse at its clock input changes from 0 to 1.

f) The first flip-flop divides the number of pulses by 2, and the second by 4.
g) This type of counter is called a ripple counter because each stage responds only after the previous stage has completed its transition.
h) The counter becomes ineffective when the counter time delay is greater than the interval between pulses.
i) In a synchronous counter, all the flip-flops are clocked synchronously by the input pulses.
j) In a synchronous counter, each stage toggles only on clock pulses that occur when the outputs of all less significant stages are 1.

3.

a) Counters usually use the binary system, but the binary output can be easily converted into decimal or other form by using suitable converting circuits.
b) An asynchronous counter consists of a series of flip-flops, the output Q of each flip-flop being connected to the clock input of the next flip-flop. The J and K inputs of all the flip-flops are connected to the supply voltage so that J = K = 1.
c) The first flip-flop divides the number of pulses at the input by 2, the second by 4 and the nth flip-flop by 2^n. A counter with n flip-flops will revert to its original state after a count of 2^n.
d) The pulses which are to be counted are applied to the clock input of the first flip-flop. All outputs are initially set to zero using the common reset line.
e) In a ripple counter, the time required for it to respond to an input pulse is approximately equal to the propagation delay time of all the flip-flops. In a synchronous counter, all the flip-flops are clocked simultaneously and the counter delay time is considerably reduced.

Exercises XXII

1.

a) Some of the tasks performed by a computer memory are, storage of the program, storage of the input data, and storage of the processed data.
b) A computer memory consists of thousands of registers in which each register can store a word.
c) A ROM has a group of registers in each of which a word is stored permanently or semipermanently.
d) Reading a memory refers to the process of making the data stored in the memory to appear at its output terminals.
e) A word stored in a memory has two parameters, its address or memory location, and the data contained in the word.
f) In a ROM, the data stored in each location is fixed at the time of manufacture. In a PROM the user himself can program the storage of data in the memory locations.
g) In a ROM, the data in each location is stored permanently, while in a RAM data can be stored and removed as required from any chosen location.
h) A static RAM uses latches as storage devices, while a dynamic RAM uses capacitors as storage devices.
i) A dynamic RAM has less transistors per memory cell, and hence allows more memory cells to be packed into an IC chip of a given size.
j) The disadvantage of a dynamic RAM is the fact that the charges on its capacitors leak away with time. Additional circuitry is required to refresh the data periodically.

2.

a) The memory of a computer stores the program, the input data, and also the processed data.
b) A ROM has a group of registers or memory locations in each of which a word is stored permanently or semipermanently.
c) The term reading a memory refers to the process of making the data stored in the memory to appear at the output terminals.

d) It is possible to read the word stored in any memory location by applying suitable electrical signals.
e) Each word stored in the memory has two parameters, its address and the data contained in the word.
f) A RAM is called a read/write memory in contrast to a ROM which is called a read only memory.
g) A PROM allows the user to store the data by using an instrument called a PROM programmer.
h) In a RAM, digital data can be stored or removed from any chosen location.
i) Semiconductor RAMs are volatile, meaning that all stored information is lost when a power failure takes place.
j) The low power consumption of CMOS memory cells makes the use of stand by battery power economical in certain applications.

3.

a) In computer, control, and information systems, it is necessary to store information and retrieve it when required.The memory is one of the most active parts of a computer, storing the program, the input data, and the processed data, at various stages of the computing process.
b) A ROM has a group of registers or memory locations in each of which a word is stored permanently. By applying suitable control signals, it is possible to read the word stored in any memory location.
c) In a ROM, the data stored in each memory location is fixed at the time of manufacture. Mass produced ROMs are usually manufactured in IC form by the IC manufacturer and not by the user.
d) Random access memories are almost invariably made in the form of IC chips. Semiconductor random access memories are volatile meaning that all stored information is lost, when a power failure takes place.
e) The storage element in a dynamic RAM is a capacitor, and a transistor is used as a transmission gate to charge the capacitor. The disadvantage of the dynamic RAM cell is that the capacitor loses its charge due to leakage currents.

Vocabulary

ability	Fähigkeit *f*
absolute zero	absoluter Nullpunkt *m*
accelerate	beschleunigen *v*
accomplish	erfüllen *v*
achieve	erreichen *v*
acquire	erwerben *v*
across	über *pr*
action	Wirkung, Handlung *f*
active filter	aktiver Filter *m*
address	Addresse *f*
adjacent	angrenzend *adj*
adjust	einstellen, regulieren *v*
advantage	Vorteil *m*
allowed	erlaubt *adj*
ambiguous	zweideutig, unklar *adj*
amplifier	Verstärker *m*
amplifier stage	Verstärkerstufe *f*
analog system	Analogsystem *n*
analysis	Analyse, Zerlegung *f*
application	Anwendung *f*
appreciable	merklich, beträchtlich *adj*
appropriate	passend, angemessen *adj*
approximately	ungefähr *adv*
arrangement	Anordnung *f*
assess	einschätzen, bewerten *v*
assign	zuordnen *v*
assume	annehmen *v*
asynchronous	asynchron *adj*
attach	anbringen, befestigen *v*
attenuator	Dämpfungsglied *n*
attractive	anziehend, reizvoll *adj*
availability	Verfügbarkeit *f*
available	vorhanden, erhältlich *adj*
avalanche	Lawine *f*
avalanche breakdown	Lawinendurchbruch *m*
avoid	vermeiden *v*
balance	ausgleichen *v*
band structure	Bandstruktur *f*
bandpass filter	Bandfilter *n*, Bandpaß *m*
bandwidth	Bandbreite *f*
bar	Stab *m*
base	Basis *f*
based on	gegründet auf
basic	grundlegend *adj*
basis	Fundament *n*, Basis *f*
behave	verhalten, sich benehmen *v*
behaviour	Verhalten, Benehmen *n*
bidirectional	in zwei Richtungen *adj*
binary device	Binärbauelement *n*
binary digit	Binärzeichen, Bit *n*
binary information	Binärinformation *f*
bipolar	bipolar *adj*
bistable multivibrator	bistabile Kippschaltung *f*
block schematic diagram	Blockschaltbild *n*
blocked state	Sperrzustand *m*

Boolean algebra	Boolesche Algebra *f*
break-over voltage	Kippspannung *f*
build	aufbauen *v*
bypass capacitor	Ableitkondensator *m*
capacitance	Kapazität *f*
cascade	Kaskadenschaltung *f*
cause	verursachen *v*
ceramic	Keramik *f*
characteristic curve	Kennlinie *f*
choice	Wahl *f*
circuit	Schaltung *f*
circulate	kreisen, umlaufen *v*
classify	klassifizieren, einteilen *v*
clear	klar *adj*
clear input	Löscheingang *m*
clock pulse	Taktimpuls *m*
CMRR	Gleichtaktunterdrükkung *f*
collector	Kollektor *m*
combination	Verknüpfung, Kombination *f*
combine	verbinden *v*
common base	Basisschaltung *f*
common collector	Kollektorschaltung *f*
common emitter	Emitterschaltung *f*
common terminal	Massenanschlußpunkt *m*
compact	kompakt, gedrängt *adj*
comparison	Vergleich *m*
compensate	ausgleichen *v*
complement	Komplement *n*, Ergänzung *f*
completely	vollständig *adv*
completely filled band	vollbesetztes Band *n*
complex numbers	komplexe Zahlen *n*
complicated	kompliziert *adj*
to be composed of	bestehen aus *v*
condition	Bedingung, Voraussetzung *f*
conduction angle	Phasenanschnittswinkel *m*
conduction band	Leitungsband *n*
conductor	Leiter *m*
configuration	Anordnung *f*
confirm	bestätigen *adj*
connection	Verbindung *f*
consequence	Folge, Wirkung *f*
consider	bedenken *v*
considerably	wesentlich, beträchtlich *adj*
constitute	bilden *v*
construction	Aufbau *m*
contact	Anschluß, Kontakt *m*
contain	enthalten *v*
content	Inhalt *m*
contrast	Gegensatz, Kontrast *m*
contribution	Beitrag *n*
control	beherrschen, kontrollieren *v*
control electrode	Steuerelektrode *f*
control system	Steuer-, Regelsystem *n*
controllable	kontrollierbar, regulierbar *v*
conversely	umgekehrt *adv*
convert	umwandeln *v*
converter	Umwandler *m*
correspond	entsprechen v
correspond to	beziehen auf, passen zu *v*
counter	Zähler *m*
coupling capacitor	Koppelkondensator *m*
covalent	kovalent *adj*
create	erzeugen *v*

criterion	Kriterium *n*
cross	kreuzen *v*
cross-couple	querverkoppeln *v*
current	Strom *m*
current gain	Stromverstärkung *f*
curve	Kurve *f*
cut-off	abschalten, abschneiden *v*
D.C (direct current)	Gleichstrom *m*
damage	Schaden *m*
damping	Dämpfung *f*
data processing system	Datenverarbeitungssystem *n*
data sheet	Datenblatt *n*
decade	Dekade *f*
decade counter	Dekadenzähler *m*
decoder	Dekodierer *m*
decrease	verkleinern, abnehmen *v*
define	definieren *v*
definite value	bestimmter Wert *m*
delay device	Verzögerungsbauelement *m*
delay equalizer	Laufzeitentzerrer *m*
denominator	Nenner *m*
density gradient	Dichtegradient *m*
depend	abhängen v
depletion mode operation	Verarmungsbetrieb *m*
depletion MOSFET	Verarmungs-MOSFET *m*
depletion region	Raumladungszone *f*
derivative	Differentialquotient *m*
describe	beschreiben *v*
design	entwerfen *v*
desired shape	gewünschte Form *f*
determine	bestimmen *v*
develop	entwickeln *v*
device	Bauelement *n*
diac	Diac *m*
different	anders, verschieden *adj*
differential input	Differenzeingang *f*
diffusion	Diffusion *f*
digital circuit	Digitalschaltung *f*
digital communication system	digitales Nachrichtensystem *n*
dimension	Dimension *f*, Maß *n*
direction	Richtung *f*
disable	sperren *v*
disadvantage	Nachteil *m*
discrete	einzeln, diskret *adj*
discuss	diskutieren, besprechen *v*
distinguish	sich unterscheiden *v*
distortion	Verzerrung *f*
distributor	Verteiler *m*
disturb	stören *v*
divider	Teiler *m*, Dividierer *m*
donor atom	Donatoratom *n*
doping	dotieren *v*
dotted	gestrichelt *adj*
drain	Drain *m*
drift	sich verschieben, treiben *v*
drift voltage	Driftspannung *f*
drive	treiben *v*
droop	Abfall *m*, Abweichung *f*
dynamic RAM	dynamisches RAM
ease	Bequemlichkeit *f*
efficient	leistungsfähig, wirksam *adj*
electrode	Elektrode *f*
elevate	erhöhen *v*
eliminate	beseitigen *v*
emitter	emitter *m*
emitter follower	Emitterfolger *m*
enable	ermöglichen, in Betrieb setzen *v*

enable	aktivieren *v*	**fanout**	Mehrfachschnittstelle *f*
energy bandgap	Energiebandabstand *m*	**fashion**	Mode *f*
enhancement mode operation	Anreicherungsbetrieb *m*	**feedback**	Rückkopplung *f*
		field	Feld *n*
enhancement MOSFET	Anreicherungs-MOSFET *m*	**field effect transistor**	Feldeffekttransistor *m*
ensure	sicherstellen *v*	**field intensity**	Feldstärke f
EPROM	löschbarer PROM *m*	**firing voltage**	Zündspannung *f*
equality detector	Äquivalenzschaltung *f*	**flat-topped characteristic**	Rechteckkurve *f*
equilibrium	Gleichgewicht *n*		
equivalent	gleichwertig *adj*	**flip-flop**	Flip-Flop, bistabiles Kippglied *n*
equivalent circuit	Ersatzschaltung *f*		
error voltage	Fehlerspannung *f*	**flywheel**	Schwunggrad *n*
establish	errichten, etablieren, herstellen *v*	**forbidden band**	verbotenes Band *n*
		forward bias	Vorwärtsvorspannung *f*
exact	genau adj	**forward current**	Durchlaßstrom *m*
exactly	genau *adv*	**four terminal network**	Vierpol *m*
example	Beispiel *n*		
exceed	überschreiten *v*	**fraction**	Bruchteil *m*
excite	anregen, erregen *v*	**frequency component**	Frequenzkomponente *f*
exhibit	ausstellen, aufweisen *v*		
exist	existieren, vorhanden sein *v*	**frequency response**	Frequenzgang *m*
		frequency selective network	frequenzselektive Schaltung *f*
expense	Aufwand *m*		
expensive	teuer *adj*	**frequently**	häufig *adv*
experience	Erfahrung *f*	**gain-bandwidth product**	Verstärkungsbandbreiteprodukt *n*
explain	erklären *v*		
expression	Ausdruck *m*	**ganged capacitor**	Mehrfachdrehkondensator *m*
extend	ausdehnen *v*		
external contact	Außenanschluß *m*	**gate**	Gate (-Anschluß) *n*
external voltage	äußere Spannung *f*	**generate**	erzeugen *v*
extract	herausziehen, extrahieren *v*	**gradual**	allmählich *adj*
		group	Gruppe *f*
extrinsic semiconductor	dotierter Halbleiter *m*	**hence**	daraus, daher *adv*
		highpass circuit	Hochpaß Schaltung *f*
fabrication	Herstellung *f*	**hole or vacancy**	Loch, Defektelektron *n*
factor	Faktor *m*, Element *n*	**hum**	Brummen *n*

hybrid equivalent circuit	Hybridersatzschaltbild *f*
hybrid parameter	Hybridparameter *m*
hybrid-π circuit	Hybridschaltung *f*
identical	identisch *adj*
identify	erkennen *v*
ignition spark	Zündfunke *m*
imbalance	Ungleichgewicht *n*
impedance transformer	Impedanzwandler *m*
implement	ausführen, in Kraft setzen *v*
imply	bedeuten *v*
improve	verbessern *v*
impurity	Verunreinigung *f*
in accordance with	in Übereinstimmung mit *adv*
in practice	in der Praxis *f*
inactive	untätig *adj*
include	enthalten, einschließen *v*
increase	zunehmen *v*
incremental	differentiell, zuwachsend *adj*
indefinitely	unbegrenzt *adj*
independent	unabhängig *adj*
indicate	anzeigen *v*
indistinguishable	unmerklich *adj*
induced channel	induzierter Kanal *m*
ineffective	wirkungslos, unwirksam *adj*
inexpensive	preiswert *adj*
influence	Einfluß *m*
initially	zuerst, anfänglich adv
input characteristic	Eingangskennlinie *f*
input impedance	Eingangsimpedanz *f*
instantaneous	sofort *adj*
insulated gate FET	FET mit isoliertem Gate *m*
insulator	Isolator *m*
interface	anschließen *v*, Schnittstelle *f*
interfere	stören, einmischen *adj*
intermetallic compound	intermetallische Verbindung *f*
intrinsic semiconductor	Eigenhalbleiter *m*
invariably	ausnahmslos *adv*
inverse-parallel	umgekehrt-parallel *adj*
invert	umkehren *v*
inverting amplifier	invertierender Verstärker *m*
inverting terminal	invertierender Eingang *m*
involve	betreffen, umfassen *v*
ionize	ionisieren *v*
junction	Übergang *m*
junction diode	Flächendiode *f*
junction FET	Sperrschicht-FET
lack of	Mangel an *n*
latch	selbsthaltender Schalter *m*
layer	Schicht *f*
lead	Anschluß *m*, Zuleitung *f*
least significant bit (LSB)	geringstwertiges Bit n
limitation	Begrenzung *f*
linearity	Linearität *f*
linearly	linear *adv*
load	Last *f*
load line	Belastungskennlinie *f*
load resistor	Lastwiderstand *m*
location	Stelle, Lage *f*
logic gate	logisches Gatter *n*
logically equivalent	logisch gleichwertig *adj*

magnitude	Größe *f*
majority carrier	Majoritätsladungsträger *m*
manufacturer	Hersteller *m*
master-slave flip-flop	bistabiles Master-Slave Kippglied *n*
match	anpassen, übereinstimmen *v*
material	Material *n*, Werkstoff *m*
mechanism	Mechanismus *m*
memory	Speicher *m*
memory element	Signalspeicher *m*
mention	erwähnen *v*
metal	Metall *n*
minority carrier	Minoritätsladungsträger *m*
mobility	Beweglichkeit *f*
modify	modifizieren, ändern, umwandeln *v*
monitoring meter	Überwachungsmeßgerät *n*
monostable oscillator	monostabile Kippschaltung *f*
most significant bit (MSB)	höchstwertiges Bit *n*
movement	Bewegung *f*
multiple	mehrfach *adj*
multiplier	Multiplizierer *m*
n-type semiconductor	n-Halbleiter *m*
narrow	eng *adj*
necessary	notwendig *adj*
negligible	vernachlässigbar *adj*
neighbouring atom	Nachbaratom *n*
noise	Geräusch *n*
noisy	rauschend *adj*
nonlinear	nichtlinear *adj*

notation	Darstellung *f*
obtain	erhalten *v*
occupy	besetzen *v*
occur	vorkommen, eintreten *v*
offset voltage	Gegenspannung *f*,
ohmic	ohmsch *adj*
omit	weglassen, übergehen *v*
open circuit	Leerlauf *m*
operating condition	Betriebsbedingung *f*
operating point	Arbeitspunkt *m*
operation	Vorgang *m*, Operation *f*
operational amplifier	Operationsverstärker *m*
optimize	optimieren *v*
oscillation	Schwingung *f*
oscillator	Oszillator *m*
output characteristic	Ausgangskennlinie *f*
output impedance	Ausgangsimpedanz *f*
overall	gesamt *adj*
overlap	überlappen *v*
p-type semiconductor	p-Halbleiter *m*
package	(Ver-) Packung *f*
parameter	Kenngröße *f*, Parameter *m*
partially filled band	teilbesetzes Band *n*
particular	besonder, speziell *adj*
peak value	Scheitelwert *m*
perform	leisten *v*
performance	Leistungsfähigkeit *f*
phase	Phase *f*
phase shift	Phasenverschiebung *f*
phase-splitter	Phasenteiler *m*
pick up	aufnehmen *v*
piezoelectric	piezoelektrisch *adj*
pinch-off	Abschnürung *f*

plentiful	im Überfluß *adj*
point of view	Standpunkt *m*
polarity	Polarität *f*
polycrystalline	mehrkristallin *adj*
popular	beliebt *adj*
popularity	Beliebtheit *f*
possess	besitzen *v*
possible	möglich *adj*
potential barrier	Potentialbarriere *f*
potential difference	Potentialdifferenz *f*
power control	Leistungsregelung *f*
power dissipation	Leistungsaufnahme *m*
power gain	Leistungsverstärkung *f*
practical	praktisch *adj*
preceding	vorhergehend *adj*
preserve	bewahren *v*
preset input	Vorwahleingang *m*
process	Verfahren *n*
produce	erzeugen *v*
progressively	fortlaufend *adj*
PROM	programmierbarer Festwertspeicher *m*
propagation delay time	Laufzeitverzögerung *f*
pulse	Puls *m*
pure state	reiner Zustand *m*
purpose	Zweck *f*
RAM	Direktzugriffspeicher *m*
range	Bereich *m*
ratio	Verhältnis *n*
read/write memory	Lese-/Schreibspeicher *m*
reason	Grund m, Ursache *f*
reasonable	vernünftig *adj*
recognize	erkennen *v*
recombination	Wiedervereinigung *f*
rectifier	Gleichrichter *m*

rectify	gleichrichten *v*
reduce	verkleinern *v*
region	Bereich *m*, Gebiet *n*
regular	regelmaßig, normal *adj*
relation	Beziehung *f*
reliability	Zuverlässigkeit *f*
reliable	zuverlässig *adj*
repetition	Wiederholung *f*
replace	ersetzen *v*
replica	Kopie, Nachbildung *f*
represent	vertreten, darstellen *v*
require	brauchen, erfordern *v*
reset line	Rücksetzleitung *f*
resistor, resistance	Widerstand *m*
respond	reagieren *v*
restrict	beschränken *v*
restricted	begrenzt *adj*
restrictive	einschränkend *adj*
result from	sich ergeben *v*
retrieve	wiederherstellen *v*
reverse bias	Sperrvorspannung *f*
revert	zurückkehren *v*
ripple counter	asynchroner Zähler *m*
rise time	Anstiegszeit *f*
role	Rolle *f*
sag	Senkung *f*, Durchhang *m*
satisfy	genügen, befriedigen *v*
saturated	gesättigt *adj*
saturation	Sättigung *f*
saturation region	Sättigungsbereich *m*
select	auswählen *v*
semiconductor	Halbleiter *m*
sensor	Meßfühler, Sensor *m*
separate	trennen *v*
separate	einzeln, getrennt *adj*
sequence	Folge *f*
sequentially	nacheinander *adv*

serial	reihenweise *adj*
shape	Form *f*
shift register	Schieberegister *n*
short circuit	Kurzschluß *m*
shunt	parallelschalten *v*
signal generator	Meßsender *m*
silicon controlled rectifier	regelbarer Siliziumgleichrichter, Thyristor *m*
simplify	vereinfachen *v*
simulate	nachbilden, simulieren *v*
simultaneously	gleichzeitig *adv*
sinusoidal	sinusförmig *adj*
slightly	ein wenig, geringfügig *adv*
small signal parameter	Kleinsignal-Parameter *m*
source	Source *f*
space	Raum *m*
space charge region	Raumladungszone *f*
specific	spezifisch, bestimmt *adj*
specification	Richtlinie, Vorschrift *f*
speed control	Drehzahlregelung *f*
stability	Stabilität *f*
stabilize	konstant halten, stabilisieren *v*
stable state	Ruhezustand *m*
stage	Stufe *f*
staggered	versetzt *adj*
step	Schritt *m*
step response	Sprungantwort *f*
storage time	Speicherzeit *f*
store	speichern, lagern *v*
subscript	Index *m*
successive	aufeinanderfolgend *adj*
suitable	geeignet, passend *adj*
supply voltage	Betriebspannung *f*
switch	schalten *v*
symmetrical	symmetrisch *adj*
synchronize	synchronisieren *v*
tangential	tangential *adj*
therefore	deshalb *adv*
thermistor	Heißleiter *m*
thyristor	Thyristor *m*
time delay	Zeitverzögerung *f*
timebase generator	Zeitablenkung *f*
toggle switch	Kippschalter *m*
train of pulses	Folge von Impulsen *f*
transfer	übertragen *v*
transition	Übergang *m*
transmission gate	Steuergatter, Durchlaßgatter *n*
transmission line	Übertragungsleitung *f*
trend	Richtung *f*,Trend *m*
triac	Triac, Zweirichtungsthyristor *m*
truth table	Wahrheitstabelle *f*
typical value	üblicher Wert *m*
unipolar	einpolig, unipolar *adj*
unique	einmalig, einzig *adj*
unity gain	Verstärkung von Eins *f*
unstable	instabil *adj*
valence band	Valenzband *n*
valence electron	Valenzelektron *n*
value	Wert *m*
varactor diode	Kapazitätsdiode *f*
variable	Variable *f*
variable capacitor	Drehkondensator *m*
variation	Änderung *f*
various	mehrere, verschiedene *adj*
versatility	vielseitigkeit *f*
vibrate	schwingen *v*
vicinity	(nähere) Umgebung *f*

video amplifier	Videoverstärker *m*	**weld**	schweißen *v*
virtually	praktisch, eigentlich *adv*	**wideband amplifier**	Breitbandverstärker *m*
volatile	energieabhängig, flüchtig *adj*	**widely**	weit *adv*
		width	Breite *f*
voltage gain	Spannungsverstärkung *f*	**work function**	Austrittsarbeit *f*
waveform generator	Wellenformgenerator *m*	**zener breakdown**	Zenerdurchbruch *m*
wavefunction	Wellenfunktion *f*		

Index

D

I

J

L

M